Advances in Intelligent Systems and Computing

Volume 899

The series "Advances in Intelligent Systems and Computing" contains publications on theory, applications, and design methods of Intelligent Systems and Intelligent Computing. Virtually all disciplines such as engineering, natural sciences, computer and information science, ICT, economics, business, e-commerce, environment, healthcare, life science are covered. The list of topics spans all the areas of modern intelligent systems and computing such as: computational intelligence, soft computing including neural networks, fuzzy systems, evolutionary computing and the fusion of these paradigms, social intelligence, ambient intelligence, computational neuroscience, artificial life, virtual worlds and society, cognitive science and systems, Perception and Vision, DNA and immune based systems, self-organizing and adaptive systems, e-Learning and teaching, human-centered and human-centric computing, recommender systems, intelligent control, robotics and mechatronics including human-machine teaming, knowledge-based paradigms, learning paradigms, machine ethics, intelligent data analysis, knowledge management, intelligent agents, intelligent decision making and support, intelligent network security, trust management, interactive entertainment, Web intelligence and multimedia.

The publications within "Advances in Intelligent Systems and Computing" are primarily proceedings of important conferences, symposia and congresses. They cover significant recent developments in the field, both of a foundational and applicable character. An important characteristic feature of the series is the short publication time and world-wide distribution. This permits a rapid and broad dissemination of research results.

**** Indexing: The books of this series are submitted to ISI Proceedings, EI-Compendex, DBLP, SCOPUS, Google Scholar and Springerlink ****

More information about this series at http://www.springer.com/series/11156

Wookey Lee · Carson K. Leung ·
Aziz Nasridinov
Editors

Big Data Analyses, Services, and Smart Data

Springer

Editors
Wookey Lee
Department of Industrial Engineering
Inha University
Incheon, Korea (Republic of)

Carson K. Leung
Department of Computer Science
University of Manitoba
Winnipeg, MB, Canada

Aziz Nasridinov
Department of Computer Science
Chungbuk National University
Cheongju, Korea (Republic of)

ISSN 2194-5357 ISSN 2194-5365 (electronic)
Advances in Intelligent Systems and Computing
ISBN 978-981-15-8730-6 ISBN 978-981-15-8731-3 (eBook)
https://doi.org/10.1007/978-981-15-8731-3

This Springer imprint is published by the registered company Springer Nature Singapore Pte Ltd.
The registered company address is: 152 Beach Road, #21-01/04 Gateway East, Singapore 189721, Singapore

Preface

Nowadays, large volumes of valuable data, which may be of different levels of veracity, can be easily generated and collected at a high velocity from a wide variety of data sources in various real-life applications and services. These big data have become a core technology to provide innovative solutions in many fields such as health care, manufacturing, and social life. Moreover, big data, big data analyses, big data applications, big data services, smart computing, and smart data have become emerging research fields that have recently drawn much attention from computer science and information technology as well as from social sciences and other disciplines.

The current volume focuses on big data analyses, services, and smart data. It consists of the following:

- Invited papers;
- Selected papers from the Sixth International Conference on Big Data Applications and Services (BigDAS 2018); and
- Extended papers from the Sixth IEEE International Conference on Big Data and Smart Computing (IEEE BigComp 2019).

The *International Conference on Big Data Applications and Services* (*BigDAS*) aims to address the need of the academic community and industry about big data. It encourages academic and industrial interaction and promotes collaborative research in big data applications and services by bringing together academics, government, and industry professionals to discuss recent progress and challenges in big data applications and services. Moreover, BigDAS also serves as a platform for theoreticians and practitioners to exchange their original research ideas on academic or industrial aspects of big data applications and services, present their new findings or innovative results on theoretical or practical aspects of big data, share their experiences on integrating new technologies into products and applications, discuss their work on performing big data applications and services in real-life situations, describe their development and operations of challenging big data-related systems, and identify unsolved challenges.

The BigDAS 2018 was the sixth edition of the conference held on August 19–22, 2018, in Zhengzhou, China, with successful previous events listed below:

- BigDAS 2015—held in conjunction with the 2015 International Conference on Digital Information Management (ICDIM 2015)—on October 20–23, 2015, in Jeju Island, South Korea;
- BigDAS 2016 on January 22–27, 2016, in Phnom Penh, Cambodia;
- BigDAS-L 2016 on December 20–23, 2016, in Vientiane, Laos;
- BigDAS 2017 on August 15–18, 2017, in Tashkent, Uzbekistan; and
- 5th BigDAS—held as part of the Asia Data Week (ADW) 2017—on November 23–25, 2017, in Jeju Island, South Korea.

BigDAS 2018 was hosted by Korea Big Data Service Society, which organized the conference together with Korea China Yeouido Leaders Forum and TusStar. General Chairs (Joo-Yeoun Lee, Hye-Kyung Kim, Shou-zhi Yu, Zhang Xu, & Wenxue Zhu) and Program Chairs (Wookey Lee, Carson K. Leung, Yan Bo Hui, & Hae-Jung Kim) of BigDAS 2018 invited authors to submit original research papers on big data applications and services, with focuses on various topics such as big data analytics, big data applications, big data frameworks, big data in business, big data in health care, big data in industry, big data models, big data security, and big data visualization. At the end, BigDAS 2018 attracted many submissions, from which the Program Committee selected 48 conference papers and 39 posters. Out of these papers, a few high-quality papers were selected to be included in the current volume.

For BigDAS 2018, the General Chairs and Program Chairs, together with the Organizing Committee—including Organizing Chairs (Kwan-Hee Yoo, Wan-Sup Cho, Yoo-Sung Kim, Che Zhong Rui, & Man-Ki Chung), Publicity Chairs (Sang-Hyun Choi, Seon-Phill Jeong, & Mun Gul), Industry Chairs (Tae-Sung Kim, Soo-young Chae, Sang-Hyun Lee, Qianjin Lai, & Man-Kyo Seo), Industry Liaison Chairs (Duck-Keun Park, Jung-Tae Kang, & Jong-Sul Chae), Demo/Poster Chairs (Su-Young Chi, Nak-hoon Baek, Jo Hyeon, & Tan Wei), Registration Chairs (Young-Ho Park, Yong-Ik Yoon Lie, & Chang Lin), Tutorial Chairs (Jin-Ho Kim & Supratip Ghose), Web Chairs (Young-Koo Lee & Jin-young Choi), Local Arrangement Chairs (Aziz Nasridinov & Yong-Bong Kim), International Advisors (Jong-beom Lee, Tae-Kyung Kim, & Shen Jie Xia), Special Session Chairs (Eun-mi Choi, Takeshi Kurata, Jae-Il Park, Young-Kyum Kim, & Kyu-Tae Lee), and Sponsor Chairs (Sung-joon Yoo, Jong-Seok Myong, Myong-Suk Jung, Jung-Min Yun, Kwang-Soo Lee, Sang-Yeub Kim, Yun-Jae Lee, Dae-Young Kim, & Ho-Kun Chae)—produced a useful and inspiring conference program with the theme of "power to change the world, big data." It consists of eight paper sessions and four poster sessions for the presentation of the 48 conference papers and 39 posters, as well as the following five keynotes:

- "Towards ubiquitous big data analytics and management: research issues and challenges" by Hiroyuki Kitagawa;

- "Programmable matter: forming objects with modular robots" by Julien Bourgeois;
- "Data science for big data applications and services" by Carson K. Leung;
- "Social network analysis in big data era" by Chaokun Wang; and
- "Era utilizing explainable artificial intelligence for power and energy" by Ho-Jin Choi.

The next (i.e., seventh) edition of the conference—BigDAS 2019—is scheduled to be held on August 21–24, 2019, in Jeju Island, South Korea. It will be organized and sponsored by Korea Big Data Service Society, Jeju National University—Software Convergence Education Center, and Chungbak National University (CBNU)—Big Data Research Institute. It will comprise three keynote speeches (by Carson K. Leung, Erel Rosenberg, and Yeong-il Seo), six technical and two poster sessions, as well as five colocated workshops.

Like BigDAS, the *International Conference on Big Data and Smart Computing* (*BigComp*) also aims to provide an international forum for exchanging ideas and information on current studies, challenges, research results, system developments, and practical experiences in the emerging fields of computer science and information technology, as well as social sciences and other disciplines. The conference was initiated by the Korean Institute of Information Scientists and Engineers (KIISE).

The IEEE BigComp 2019 was the sixth edition of the conference held on February 27–March 2, 2019, in Kyoto, Japan, with successful previous events listed below:

- BigComp 2014 on January 15–17, 2014, in Bangkok, Thailand;
- BigComp 2015 on February 09–12, 2015, in Jeju Island, South Korea;
- BigComp 2016 on January 18–20, 2016, in Hong Kong, China;
- IEEE BigComp 2017 on February 13–16, 2017, in Jeju Island, South Korea; and
- IEEE BigComp 2018 on January 15–18, 2018, in Shanghai, China.

IEEE BigComp 2019 was co-sponsored by the IEEE, IEEE Computer Society and KIISE, in cooperation with the Database Society of Japan (DBSJ). General Chairs (Masatoshi Yoshikawa & Jongwon Choe) and Program Chairs (Hiroyuki Kitagawa, Walid Saad, Kyuseok Shim, & Jie Tang) of BigComp 2019 invited authors to submit original research papers and original work-in-progress reports on big data and smart computing, with focuses on various topics such as big data & cloud systems, data mining, embedding & ontology, face recognition & speech, machine learning, mobility & location data, network & Internet of things (IoT), query processing, smart city, social media, spatial data & query, text & natural language, time series, traffic analysis, as well as visual data & virtual reality (VR). IEEE BigComp 2019 attracted 170 paper submissions, from which the Program Committee selected 39 regular papers and 34 short papers. These 73 conference papers, together with papers from seven workshops and abstracts of the two keynotes (by Zachary Ives & Raymond Ng), were published in an IEEE conference proceedings. Out of these papers, a few high-quality papers were invited for extension and expansion to be included in the current volume.

The next edition of the conference—IEEE BigComp 2020—is scheduled to be held on February 19–22, 2020, in Busan, South Korea. It will comprise three keynote speeches (by Masaru Kitsuregawa, Jiawei Han, and Sung-Bae Cho), several technical sessions, tutorials, and workshops.

As for the current volume, it would not have been possible without the help and effort of many people and organizations. We express our thanks to the Organizing Committee, Program Committee, hosts, and sponsors of BigDAS 2018 and IEEE BigComp 2019. We thank authors and non-author participants of both conferences. We are grateful to reviewers for their professionalism and dedication in different aspects, especially in the selection of papers presented in the current volume. Last but not least, we thank the staff at Springer (especially, Smith Chae, Eugene Hong, Annie Kang, Prashanth Ravichandran) for their help in publishing the current volume.

July 2019

Wookey Lee
Carson K. Leung
Aziz Nasridinov

Contents

Analysis of the Effects of Nature and Facility Environmental Attributes on the Cause of Death from Disease

Kyoung-ae Jang and Woo-Je Kim(✉)

Department of Industry and Information Systems Engineering, Frontier Laboratory Room 707, Seoul National University of Science and Technology, 232, Gongneung-ro, Nowon-gu, Seoul 01811, Republic of Korea
jkalove@hanmail.net, wjkim@seoultech.ac.kr

Abstract. In this study, we investigated the correlations between environmental factors and diseases related to death in Korea. We collected data on natural harmful and social environmental factors from 2015 to 2017 in Korea. We defined the cause of death by disease as a dependent variable and conducted analyses to derive related attributes. We used a multiple linear regression analysis and classification and regression tree models as analytical methods to derive associated attributes. The results of this study indicate that natural environmental factors have a negative correlation with the cause of death by disease, while harmful materials and facilities have positive correlations. Through this study, we hope that people will recognize the importance of protecting the natural environment and the risks of harmful substances.

Keywords: Cause of death · Multi linear regression · Classification and regression tree analysis · Big data

1 Introduction

As interest in health and disease grows, the number of studies on environmental factors affecting health and disease is also increasing. The Organization for Economic Co-operation and Development (OECD) "economic consequences of outdoor air pollution" announcement states that the number of premature deaths due to air pollution around the world is expected to increase from 6 million to 9 million by 2060 [1]. In Korea, the Ministry of the Environment reported 258 days with bad fine dust level air pollution in 2016 and suggested comprehensive measures to minimize the damage to public health [2]. The OECD also states that contaminated drinking water and other environmental pollution sources threaten health and increase the number of premature deaths [3].

Many previous studies have investigated health determinants and causes of death by identifying a number of different factors depending on social changes in the regions and have highlighted the need for national and periodic studies. There have been studies on health-related factors such as health inequality [4], health and facility environment [5–7], health and socioeconomic factors [8–11], and so on. There have also been studies

W. Lee et al. (Eds.): *Big Data Analyses, Services, and Smart Data*, AISC 899, pp. 1–14, 2021.
https://doi.org/10.1007/978-981-15-8731-3_1

on the importance of the facility environment for health such as on the development of the walking environment [5], the importance of city planning policy for citizens' health [6], and to propose a health framework for urban-dwelling people [7]. However, it has been difficult to find studies on the relationship between the natural environment and pollution by harmful materials and the cause of death by disease. Additionally, it has also been difficult to find research on the effect of natural environmental factors and harmful environmental factors on human health.

Therefore, the purpose of this study was to investigate the correlations between environmental factors and diseases related to death in Korea. To do this, we first collected data on natural, harmful and social environmental factors from 2015 to 2017 from the Korea Statistical Information Service (KOSIS). The data comprised 1,050,700 data points for 68 attributes possibly related to the causes of death by disease. Next, we applied a multiple linear regression analysis (MLRA) and a classification and regression tree (CART) analysis, which are advantageous for linear and nonlinear regression data, respectively, to derive attributes with a high impact on the causes of death by disease. We applied these methods to combine the advantages of linear and nonlinear regression models.

The results of the analyses in this study showed that natural and harmful environmental factors affect human disease prevalence: we derived the harmful environmental properties that promote disease and the natural environment properties that mitigate disease. Thus, we confirmed the serious requirement needed to protect nature and the danger of harmful facilities. This study provides the basic background for establishing policy and decision-making for health protection, environment management, harmful facilities management, and future environmental protection.

The composition of this paper is as follows. Section 2 contains an analysis of previous studies in this research field. In Sect. 3, the research design is established and an analysis of the results is performed. Section 4 presents conclusions of the study.

2 Previous Research

2.1 Analysis of Causes of Illness and Death

Yoon and Kim [4] used a correlation analysis between health and disease to study social and economic status according to income; they showed significant health inequality between people in urban and rural areas due to the income gap. Lawrence et al. proposed that the walking environment and physical characteristics of a community can affect the health of the local population [5], while Corburn [6], and Freudenberg and Valhov [7], also showed the importance of the physical environment on human health. Other studies have found that socioeconomic factors such as income, education, and the vulnerability of healthcare also affect health and disease outcomes [8–10]. Cohen-Mansfield et al. showed that social, physical, and institutional environmental problems in residential areas were the main impacts on human health and claimed that designing pedestrian-friendly facilities helps residents to be healthy [11]. Meanwhile, Giles-Corti and Donovan analyzed interactions between the physical environment, physical activity, and health status [12].

The major causes of US deaths in 2011 were diseases of the heart, malignant neoplasms, and chronic lower respiratory diseases [13]. Mathers et al. investigated the death statistics for 115 countries and came to the conclusion that "Few countries have good-quality data on mortality that can be used to support policy development and implementation adequately" and that "System improvement was necessary" [14].

Recent research on the effects of air pollution and dust on human health has increased considerably [15–17]. Some studies have suggested that air quality is becoming worse and is increasing related illnesses and deaths as industry develops in China [15], how much indoor air affects health [16], and how causes of childhood allergic rhinitis are affected by the frequency of truck traffic near residences [17]. It has also been found that gas combustion affects health by polluting the air [18]. There has also been a research paper on the relationship between dust and death in Korea [19].

There have been studies using simple statistical facts or methods [15, 16, 18], and regression analysis methods have been widely used to derive major attributes [17, 19–21]. Although we found an analysis study using the CART method of risk factors for hospital mortality [22], there have not yet been studies on health outcomes using regression and decision tree methods. In this study, MLRA was applied as a typical regression method for linear relationships between variables, while CART was added to supplement the analysis results by creating a predictive model that can perform a nonlinear relationship analysis.

2.2 MLRA (Multiple Linear Regression Analysis)

The assumption for MLRA is that there is a linear relationship between independent variables (X_1, X_2, …, X_n) and dependent variable (Y). Therefore, we can estimate the regression coefficient (β), which is the influence of each independent variable, on the basis of the training data:

$$Y = \beta_0 + \beta_1 X_1 + \beta_2 X_2 \ldots . \beta_n X_n + \varepsilon \tag{1}$$

We could confirm the influence of multiple independent variables using MLRA to select only significant variables from candidate independent variables. The selection method used supervised variable selection techniques such as forward, backward elimination, and stepwise selection. We also applied the selection method to analyze the causal attributes of death from disease.

2.3 CART (Classification and Regression Tree)

We constructed a classification model using the CART algorithm, which is a type of decision tree method. CART uses a decision tree as a predictive model to classify conclusions regarding the target value of an item leaves from observations about it branches. This predictive modeling method is commonly used in statistics, data mining, and machine learning. A decision tree model where target variables take a discrete set of values is called a classification tree and one where the target variables are continuous is called a regression tree.

Unlike other regression analysis methods, the CART method classifies via objective rules instead of assumptions [22]. In addition, the method can overcome the overfitting and multicollinearity problems of MLRA, and it is possible to regressively analyze nonlinear data when they are categorical or continuous [23, 24]. Therefore, in this study, we used the CART method to compensate for the disadvantages of MLRA.

3 Research Design and Results

3.1 Research Design

Figure 1 shows the framework of this research. We incorporated data collection, data preprocessing, optimal variable selection, MLRA, classification and reduction using CART, and a comparison of the MLRA and CART results. Data collection included assessing outcomes of previous studies and collecting natural environment and harmful facility data from KOSIS for a correlation analysis of death by disease. We then defined independent and dependent variables and parameters in the collected data and created appropriate data metrics.

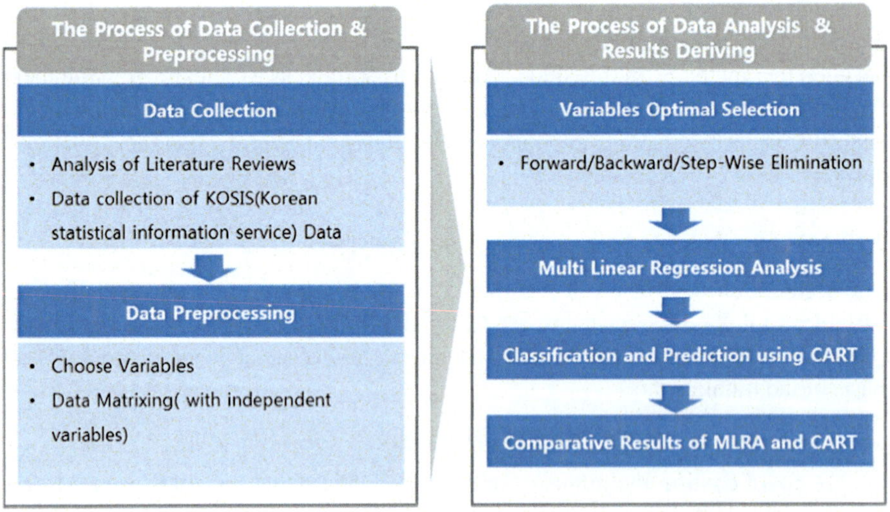

Fig. 1. Research framework

3.2 Research Results

3.2.1 Data Collection and Preprocessing

This study was conducted under the assumption that diseases could be affected by environmental factors and harmful facilities. We established four hypotheses and collected relevant attribute variables to test them.

- Hypothesis 1. Mountains, rivers, forests, and prohibited areas will have a negative correlation with disease if the area is large enough.
- Hypothesis 2. Many harmful materials and facilities including wastes, slaughterhouses, and oil supply facilities will have a positive correlation with disease.
- Hypothesis 3. The types of disease will be related to the natural environment, and harmful material and facility types.

- Hypothesis 4. Natural environment, and harmful materials and facilities will have more impact on disease than social environment.

To analyze the cause of death by disease in Korea, we collected data related to the population and society, harmful materials and facilities, and the natural environment from the data freely available from the Korea National Statistical Office.

The collected attribute data have three categories, as reported in Table 1. The data for the Population & Social area were composed of the population, moving population, gender, age-standardized mortality rate, economic activity participation rate, employment

Table 1. The collected data

Data area	Attribute
Population & society	Cities and provinces, cities & provinces and towns, total population & CP population, in transfer rate(%), out transfer rate(%), Moving Index, Population-to-Transfer Index, gender, age standardized mortality rate(%), Economic activity participation rate(%), Employment rate (%), Education index, leisure and leisure facilities using rate(%), work stress index, overall stress index, drinking index, smoking index, social support index
Harmful materials & facilities	Facilities handling air pollution (number), facilities handling hazardous materials (number), wastewater discharge (m^3/day), organic material load (kg/day), waste pesticide containers (number), waste landfill (tons/year), waste incineration (tons/year), household waste (ton/day), standard plastic garbage bag emission (tons/day), separate emissions of recyclable resources (ton/day), food emission (ton/day), construction waste (ton/ day), non-combustible waste (ton/day), flammable waste (ton/day), mixed waste (ton/day), other waste (ton/day), site waste (ton/day), combustible waste (ton/day), noncombustible waste (ton/day), chemical substances emission (kg/year), carcinogenic concern emissions (kg/year), cattle (number)
Natural environment	Forest area (ha), factory area (m^2), space facility area (m^2), airport area (m^2), port area (m^2), total area (m^2), conservation area (m^2), river area (m^2), reservoir area (m^2), slaughterhouse area (m^2), crematorium area (m^2), funeral home area (m^2), comprehensive medical facility area (m^2), broadcasting communication facility area (m^2), heat supply facility area (m^2), oil supply facility area (m^2), development restricted Area (m^2), natural environment conservation area (m^2), agriculture and forest area (m^2), water pollution protection facilities area (m^2),waste treatment facility area (m^2), sewerage facility area (m^2)

rate, education index, leisure and leisure facilities use rate, work stress index, overall stress index, drinking index, smoking index, social support index, and so on.

The data on harmful facilities had 22 subcategories, including facilities handling air pollution (number), facilities handling hazardous materials (number), wastewater discharge (m^3/day), organic material load (kg/day), waste pesticide containers (number), waste landfill (tons/year), and so on.

The natural environment data also had 22 subcategories (Table 1), including forest area, forest area (ha), factory area (m^2), space facility area (m^2), airport area (m^2), port area (m^2), total area (m^2), conservation area (m^2), river area (m^2), waste incineration (tons/year), and so on.

3.2.2 Analysis of the Effects of Natural and Harmful Environment Attributes

We investigated the correlations between death by disease and the natural environment and harmful material and facility attributes, where the dependent variable comprised codes for causes of death by disease, as listed in Table 2 (the Korean standard classification codes).

Table 2. Korean standard classification code for causes of death by the disease

Class_No.	Classification code
001	Certain infectious and parasitic diseases (A00–B99)
002	Neoplasm (C00–D48)
003	Specific disorders that involve blood and hematopoietic organ disease and immune mechanisms (D50–D89)
004	Endocrine, nutritional and metabolic diseases (E00–E88)
005	Mental and behavioral disorders (F01–F99)
006	Diseases of the nervous system (G00–G98)
007	Ear and mastoid disease (H60–H93)
008	Circulatory system diseases (I00–I99)
009	Diseases of the respiratory system (J00–J98, U04)
010	Diseases of the digestive system (K00–K92)
011	Diseases of skin and subcutaneous tissue (L00–L98)
012	Diseases of musculoskeletal and connective tissue (M00–M99)
013	Diseases of the genitourinary system (N00–N98)
014	Pregnancy, Childbirth and Late Birth (O00–O99)
015	Specific conditions originating in the prenatal period (P00–P96)
016	Congenital anomalies, deformities and chromosomal abnormalities (Q00–Q99)
017	Symptoms, signs not otherwise specified (R00–R99)
018	Outsider of disease morbidity and mortality (V01–Y89)

We analyzed 22 natural environment attributes and 22 harmful material and facility attributes by applying MLRA to derive factors with a major influence on the attribute data. An MLRA model was constructed by selecting optimal variables after applying forward selection, backward elimination, and stepwise selection. We used 70% of the available data to build the models and 30% for verification.

Table 3 summarizes the MLRA results for each disease-causing variable in which all values with significance <0.1 were extracted using R.

Table 3. Results of MLRA Analysis on the natural and harmful environment attributes.

Class_No.	Variables	Estimate	Std. error	t value	Pr(>|t|)	Signif. codes
001 (A00–B99)	Carcinogenic concern emissions (kg/year)	9.32E − 06	4.15E − 06	2.245	2.48E − 02	*
	Agriculture and forest area (m^2)	−1.60E − 05	8.46E − 06	−1.888	5.91E − 02	.
002 (C00–D48)	Noncombustible waste (ton/day)	7.67E − 04	2.72E − 04	2.82	4.82E − 03	**
003 (D50–D89)	Separate emissions of Recyclable resources (ton/day)	−2.31E − 02	1.04E − 02	−2.228	2.59E − 02	*
	Waste incineration (tons/year)	9.77E − 05	4.38E − 05	2.231	0.0257	*
	Reservoir area (m^2)	−2.54E − 02	1.42E − 02	−1.791	7.34E − 02	.
004 (E00–E88)	Noncombustible waste (ton/day)	6.62E − 04	2.63E − 04	2.516	0.0119	*
005 (F01–F99)	Port area (m^2)	−3.15E − 03	1.89E − 03	−1.662	0.09656	.
	Forest area (ha)	−4.22E + 01	2.41E + 01	−1.752	0.07977	.
	Noncombustible waste (ton/day)	7.41E − 04	3.62E − 04	2.049	0.04055	*
006 (G00–G98)	Sewerage facility area (m^2)	−1.35E − 02	7.49E − 03	−1.799	0.0721	.
007 (H60–H93)	Mixed waste (ton/day)	1.69E − 03	3.42E − 04	4.923	8.84E − 07	***
	Waste treatment facility area (m^2)	8.25E − 04	1.93E − 04	4.269	2.01E − 05	***

(*continued*)

Table 3. (*continued*)

Class_No.	Variables	Estimate	Std. error	t value	Pr(>|t|)	Signif. codes
	Reservoir area (m^2)	3.20E − 03	1.23E − 03	2.608	0.00915	**
	Food emission (ton/day)	−2.51E − 03	8.76E − 04	−2.871	0.00411	**
	Funeral home area (m^2)	−7.30E − 02	2.45E − 02	−2.975	0.00295	**
	River area (m^2)	−3.72E − 05	2.05E − 05	−1.812	0.06999	.
008 (I00–I99)	Noncombustible waste (ton/day)	6.27E − 04	2.66E − 04	2.357	0.0185	*
009 (J00–J98, U04)	Carcinogenic concern emissions (kg/year)	8.36E − 06	3.78E − 06	2.21E + 00	0.0269	*
010 (K00–K92)	Chemical substances emission (kg/year)	1.47E − 06	6.88E − 07	2.141	0.0323	*
011 (L00–L98)	Separate emissions of recyclable resources (ton/day)	−8.352E − 03	4.617E − 03	−1.809	0.0705	.
012 (M00–M99)	Food emission (ton/day)	1.5513E − 02	6.325E − 03	−2.452	0.0142	*
013 (N00–N98)	Cattle (number)	4.88E − 06	2.42E − 06	2.014	0.0441	*
014 (O00–O99)	Waste treatment facility area (m^2)	4.36E − 03	7.25E − 04	6.018	1.91E − 09	***
	River area (m^2)	−1.89E − 04	6.68E − 05	−2.822	0.00479	**
	Broadcasting communication facility area (m^2)	5.50E − 02	3.12E − 02	1.761	0.07822	.
	Wastewater discharge (m^3/day)	1.21E − 04	7.31E − 05	1.653	0.0985	.
015 (P00–P96)						

(*continued*)

Table 3. (*continued*)

Class_No.	Variables	Estimate	Std. error	t value	Pr(>|t|)	Signif. codes
	Separate emissions of recyclable resources (ton/day)	−2.01E − 02	4.36E − 03	−4.606	4.22E − 06	***

Signif. codes: 0 '***' 0.001 '**' 0.01 '*' 0.05 '.' 0.1 ' ' 1

We derived properties with meaningful relationships using the MLRA model. For mortality due to disease, natural environment attributes such as the agriculture and forest area (m^2), and port, river and reservoir area (m^2) exhibited negative correlations whereas harmful material and facility attributes such as carcinogenic concern emissions (kg/year), noncombustible waste (ton/day), waste incineration (tons/year), and chemical substances emission (kg/year) exhibited positive correlations with mortality due to disease. Among them, noncombustible waste (ton/day) and carcinogenic concern emissions (kg/year) were the most influential. Therefore, the results confirm that the natural environment attributes and harmful facility attributes are related to human health.

It is confirmed that hypothesis 1–3 defined at the beginning of the research was supported. We found four anomalous attributes. Separate emissions of recyclable resources (ton/day), food emission (ton/day), sewerage facility area (m^2), and funeral home area (m^2) were negatively correlated with mortality, all of which we expected to be positively so. Thus, our interpretation is that the first two attributes reflect the cleaning up the local environment and the last one reflects good air quality. It can be interpreted as the separation and discharge of food and recycled wastes to clean the environment and to affect health.

To verify hypothesis 4 (the natural environment and harmful materials and facilities have a greater impact on disease than the social environment), we collected 19 corresponding attributes and defined them as independent variables which are population and social environment attributes (Table 1).

An additional MLRA analysis was performed to include these attributes along with the previous harmful material and facility ones, as reported in Table 4.

We derived five classifier attributes with a significant value $| P | < 0.1$ based on the P-values given in Table 4. The experimental results show that three of the population and society attributes and two of the natural and harmful environmental attributes were related to the cause of death by disease. We found two significant attributes with a significant value $| P | < 0.05$ based on the P-values: gender and smoking index of population & society area. It was not confirmed whether hypothesis 4 defined at the beginning of the research was supported.

Table 4. Results of MLRA Analysis on the population & society area.

Variables	Estimate	Std. Error	t value	Pr(>|t|)	Signif. codes
Gender	1.85E + 01	4.51E − 01	40.99	<2e − 16	***
Education index	−2.09E + 01	1.25E + 01	−1.672	0.0945	.
Carcinogenic concern emissions (kg/year)	5.12E − 04	2.67E − 04	1.915	0.055603	.
Economic activity participation rate(%)	−1.94E − 01	1.12E − 01	−1.728	0.084039	.
Smoking index	4.41E − 01	2.25E − 01	1.961	0.049955	*

Signif. codes: 0 '***' 0.001 '**' 0.01 '*' 0.05 '.' 0.1 ' ' 1

3.2.3 Comparative Experimental Results with CART

A classification model was constructed using CART to supplement the attributes that could not be derived using MLRA. The CART explanatory variable was derived from the number of waste pesticide Containers and the number of cattle, as shown in Fig. 2.

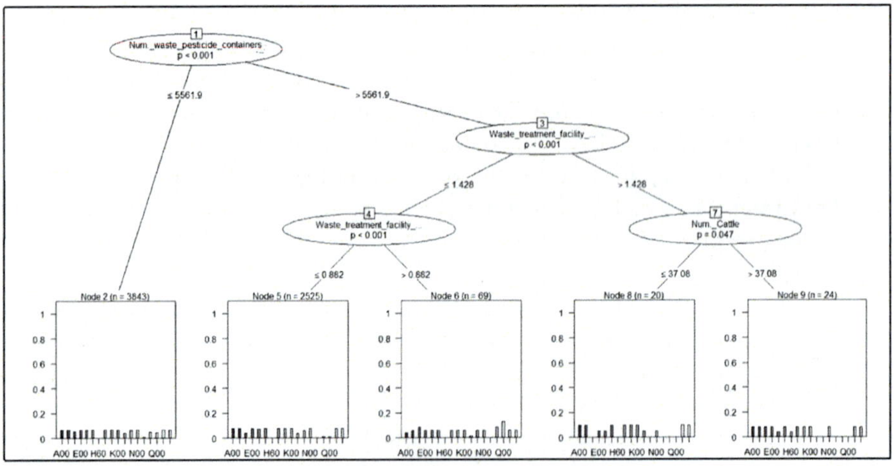

Fig. 2. The results of CART analysis

Using the CART method, we can find the attributes that classify the set and the value of the classifier well. We derived a classifier attribute with a significant value | P | < 0.05 based on the P-value with the CART method (Table 5). The tree structure of CART generates right or left child nodes if the selected attribute value is larger than, or smaller than or equal to, the reference value, respectively. Using CART confirms the model with the classifier value and the tree through this iterative process.

Table 5. Results of CART analysis on the natural and harmful environment area.

System_variables/Variables	Node name	Classifier value	Number of node	Pr(>|t|)	Signif. Codes
Num._waste_pesticide_containers/waste pesticide containers (number)	Node 2	5561.900	3843	<0.001	***
Waste_treatment_facility_./ waste treatment facility area (m^2)	Node 5	1.428	2525	<0.001	*
Waste_treatment_facility_./ waste treatment facility area (m^2)	Node 6	0.882	69	<0.001	*
Num._Cattle/ cattle (number)	Node 8	37.080	20	0.047	*
Num._Cattle/ cattle (number)	Node 9	37.080	24	0.047	*

Signif. Codes: 0 '***' 0.001 '**' 0.01 '*' 0.05 '.' 0.1 ' ' 1

The results of this experiment enabled us to classify three attributes and five groups. Node2 classification is "Num._waste_pesticide_containers" (waste pesticide containers (number)) $\leq$ 5561.9 and Node5 classification is "Num._waste_pesticide_containers"(waste pesticide containers (number)) > 5561.9 and "Waste_treatment_facility." (waste treatment facility area (m^2)) $\leq$ 0.882.

Table 6 shows the results of CART and MLRA analysis.

We defined 19 significant attributes derived from MLRA and three attributes derived from CART. There were 9 negative correlations and 10 positive ones. Negatively correlated attributes agriculture and forest area (m^2), forest area (ha), funeral home area (m^2), airport area (m^2), reservoir area (m^2), river area (m^2) were for properties related to the natural environment.

We found four anomalous attributes: Separate emissions of recyclable resources (ton/day), food emission (ton/day), sewerage facility area (m^2), and funeral home area (m^2) which were negatively correlated with mortality. We interpreted these as facilities that create or else are located in a good environment and the separation: discharge of food and recycled wastes affected clean the environment and to affect health. Positively correlated attributes (waste pesticide containers (number), waste incineration (tons/year), waste treatment facility area (m^2), wastewater discharge (m^3/day), broadcasting communication facility area (m^2), carcinogenic concern emissions (kg/year), chemical substances emission (kg/year), mixed waste (ton/day), non-combustible waste (ton/day), cattle (number)) were for properties related to harmful substances and facilities. As we experimented with CART, we were able to obtain firm results that further proved hypotheses 1–3.

4 Conclusion

Many researchers have recognized the importance of the relationship between the environment and health. However, it was difficult to find studies based on the assumption

Table 6. Results of CART and MLRA analysis

No.	Variables	Correlations	Methods	Signif. codes
1	Agriculture and forest area (m^2)	Negative	MLRA	.
2	Broadcasting communication facility area (m^2)	Positive	MLRA	.
3	Carcinogenic concern emissions (kg/year)	Positive	MLRA	*
4	Chemical substances emission (kg/year)	Positive	MLRA	*
5	Separate emissions of recyclable resources (ton/day)	Negative	MLRA	***
6	Food emission (ton/day)	Negative	MLRA	**
7	Forest area (ha)	Negative	MLRA	.
8	Funeral home area (m^2)	Negative	MLRA	**
9	Mixed waste (ton/day)	Positive	MLRA	***
10	Non-combustible waste (ton/day)	Positive	MLRA	**
11	Cattle (number)	Positive	MLRA, CART	*, *
12	Waste pesticide containers (number)	Positive	CART	*
13	Airport area (m^2)	Negative	MLRA	.
14	Reservoir area (m^2)	Negative	MLRA	**
15	River area (m^2)	Negative	MLRA	**
16	Sewerage facility area (m^2)	Negative	MLRA	.
17	Waste incineration (tons/year)	Positive	MLRA	*
18	Waste treatment facility area (m^2)	Positive	MLRA, CART	***, ***
19	Wastewater discharge (m^3/day)	Positive	MLRA	.

Signif. codes: 0 '***' 0.001 '**' 0.01 '*' 0.05 '.' 0.1 ' ' 1

that pollution of the natural environment with harmful substances is bad for health that analyzed the relationship. The originality of this study is the analysis of the causes of death due to diseases by considering the natural environmental and harmful materials and facility factors at the same time.

The purpose of this study was to investigate correlations between health and the natural environment and harmful factor attributes. In this study, we conducted analyses with MLRA and CART methods via the R programming language. MLRA has advantages for linearly regressed data, while CART is a strong method for nonlinearly regressed data. Therefore, we applied both research methods to combines the advantages of linear and nonlinear regression models.

We tested four hypotheses regarding how the natural environment and harmful factors could affect human health. We collected attribute data from the population and society, hazardous substances and facilities, and the natural environment datasets supplied by the

Korea National Statistical Office as publicly available data and constructed regression and classification models using MLRA and CART, respectively

- Hypothesis 1 was supported: the size of natural environmental areas, such as mountains, rivers, forests, and forbidden areas, exhibited negative correlations with disease mortality.
- Hypothesis 2 was supported: Hazardous materials and facilities, such as wastes, harmful facilities, and slaughterhouses, exhibited positive correlations with disease mortality.
- Hypothesis 3 was supported: there was significant correlation between disease type and property type.
- Hypothesis 4 was not supported: natural and harmful environments had less significant impact than social environment.

Thus, this study showed that the natural environment and harmful facilities are significantly related to human health. We defined 19 significant attributes with MLRA and three attributes derived from CART: nine negative correlations and 10 positive ones. The negatively correlated attributes are properties related to the natural environment, such as the agricultural and forest area, and port, river, and reservoir area, while the positive ones are properties related to harmful substances and facilities, such as carcinogenic concern, combustible waste, waste incineration, and chemical substance emission facilities.

The study was limited to specific data for Korea and was not generalized. Further studies are required to analyze various additional factors and map the cause of death due to specific diseases to the various attributes.

References

1. OECD: The economic consequences of outdoor air pollution (2017)
2. Ministry of Environment: Comprehensive measures for fine dust management (2017)
3. OECD: Health and Environment (2008)
4. Yoon, T.H., Kim, J.H.: Health inequality between urban and rural areas in South Korea. J. Korean Soc. Rural Living Nurs. **1**(1), 11–20 (2006)
5. Lawrence, J.J., Grinspan, Z.M., Statland, J.M., McBain, C.J.: Muscarinic receptor activation tunes mouse stratum oriens interneurones to amplify spike reliability. J. Physiol. **571**(3), 555–562 (2006)
6. Corburn, J.: Confronting the challenges in reconnecting urban planning and public health. Am. J. Public Health **94**, 541–546 (2004)
7. Galea, S., Freudenberg, N., Vlahov, D.: Cities and population health. Soc. Sci. Med. **60**(2005), 1017–1033 (2005)
8. Farmer, M.M., Ferraro, K.F.: Are racial disparities in health conditional on socio economic status? Soc. Sci. Med. **60**(1), 191–204 (2005)
9. Kelley-Moore, J.A., Ferraro, K.F.: The black/whiter disability gap: persistent inequality in later life? J. Gerontol. Ser. B: Psychol. Sci. Soc. Sci. **59**(1), S34–S43 (2004)
10. Williams, D.R., Sternthal, M.: Understanding racial-ethnic disparities in health: sociological contributions. J. Health Soc. Behav. **51**(1), S15–S27 (2010)
11. Cohen-Mansfield, J., Marx, M.S., Guralnik, J.M.: Motivators and barriers to exercise in and older community-dwelling population. J. Aging Phys. Act. **11**(2), 242–253 (2003)

12. Giles-Corti, B., Donovan, R.J.: The relative influence of individual, social and physical environment determinants of physical activity. Soc. Sci. Med. **54**(12), 1793–1812 (2002)
13. Kenneth, D., Kochanek, M.A.: Deaths: final data for 2011. Natl. Vital Stat. Rep. **63**(3), 1–120 (2015)
14. Mathers, C.D., Fat, D.M., Inoue, M., et al.: Counting the dead and what they died from: an assessment of the global status of cause of death data. Bull.-Word Health Organ. **83**, 171–177 (2005)
15. Chen, Z., Wang, J.N., Ma, G.X., Zhang, Y.S.: China tackles the health effects of air pollution. Lancet **382**(9909), 1959–1960 (2013)
16. Perez-Padilla, R., Schilmann, A., Riojas-Rodriguez, H.: Respiratory health effects of indoor air pollution. Int. J. Tuberc. Lung Dis. **14**(9), 1079–1086 (2010)
17. Shirinde, J., Wichmann, J., Voyi, K.: Allergic rhinitis, rhinoconjunctivitis and hayfever symptoms among children are associated with frequency of truck traffic near residences: a cross sectional study. Environ. Health **14**(1), 84 (2015)
18. Najjar, Y.S.: Gaseous pollutants formation and their harmful effects on health and environment. Innov. Energy Policies **1**, 1–9 (2011)
19. Kwon, H.J., Cho, S.H., Chun, Y., Lagarde, F., Pershagen, G.: Effects of the Asian dust events on daily mortality in Seoul, Korea. Environ. Res. **90**(1), 1–5 (2002)
20. Abdel-Shafy, H.I., Mansour, M.S.: A review on polycyclic aromatic hydrocarbons: source, environmental impact, effect on human health and remediation. Egypt. J. Petrol. **25**(1), 107–123 (2016)
21. Stuckler, D., Basu, S., Suhrcke, M., Coutts, A., McKee, M.: The public health effect of economic crises and alternative policy responses in Europe: an empirical analysis. Lancet **374**(9686), 315–323 (2009)
22. Fonarow, G.C., Adams, K.F., Abraham, W.T., Yancy, C.W., Boscardin, W.J., ADHERE Scientific Advisory Committee: Risk stratification for in-hospital mortality in acutely decompensated heart failure: classification and regression tree analysis. Jama, **293**(5), 572–580 (2005)
23. Breiman, L.: Classification and Regression Trees. Kluwer Academic Publishers, New York (1984)
24. Yohannes, Y., Hoddinott, J.: Classification and regression trees: an introduction. International Food Policy Research Institute (1999)

Big Data Computing and Mining in a Smart World

Carson K. Leung(✉)

University of Manitoba, Winnipeg, MB, Canada
kleung@cs.umanitoba.ca

Abstract. A smart world can be considered as a convergence of the physical world, cyber world, social world, and thinking world. In these four worlds, huge amounts of valuable data are generated and gathered at a rapid rate from a broad range of data sources. Although the quality of these big data depend on their degrees of uncertainty, rich sets of valuable information and useful knowledge can be mined from the big data. This paper focuses on *big data computing and mining*, which aims to (a) analyze these rich sets of big data, and (b) discover implicit, previously unknown, and potentially useful information and knowledge from the big data. In particular, we present data science solutions for discover frequent patterns. Through our presentation, we discuss how these solutions interconnecting (a) big data generated and collected from the physical world, (b) frequent pattern mining algorithms in the cyber world, (c) social interactions among social individuals in the social world, and (d) user preference and interest reflecting the user cognitive thinking in the thinking world. We show these interactions through our discussion on mining coronavirus disease 2019 (COVID-19) data in a smart world environment. The interconnections link the physical, cyber, social and thinking worlds together to establish a better environment towards big data computing and mining in a smart world.

Keywords: Cybermatics · Smart world · Big data · Data science · Data mining · Frequent patterns · Constraints · Social network analysis · Physical world · Cyber world · Social world · Thinking world · COVID-19

1 Introduction and Related Works

Recent advancements in information science and technology lead to rapid generation (e.g., streaming data) of huge amounts of valuable data (e.g., financial, medical, shopper market basket data). These big data are usually gathered at different degrees of veracities (e.g., precise data, imprecise and uncertain data) and from a broad range of data sources in numerous business, engineering, health, industrial, scientific, and social applications in a *smart world* [1, 2]. In general, a smart world can be considered as a physical-cyber-social-thinking hyperspace or converged world [3]. It consists of the following four worlds:

W. Lee et al. (Eds.): *Big Data Analyses, Services, and Smart Data*, AISC 899, pp. 15–27, 2021.
https://doi.org/10.1007/978-981-15-8731-3_2

1. *Physical world*, which is either a natural or a human-made system operating for big data applications and services. For example, ubiquitous sensors or actuators interact (e.g., perceive, control) physical objects through technologies for information, communication and networking.
2. *Cyber world*, which is a generalized information resource with discrete logical and switched computation, communication, as well as control.
3. *Social world*, which is a semantic integration of social attributes, as well as both social intra-relationships and social inter-relationships among social entities (e.g., human users, physical objects, cyber entities).
4. *Thinking world*, which is a reflection of human brain activities for cognition processes of analysis, judgment, reasoning and synthesis, based on abstract conceptions and/or concrete representations.

With these four physical-cyber-social-thinking worlds within the smart world, ubiquitous things establish the following dynamic and seamless interconnections within the smart world:

1. *Physical-based coordination*—which refers to mutual assistance relationships established through critical infrastructures, heterogeneous interfaces, and/or interactive environments—such as
   a. coordination between identifier-based and non-identifier-based objects,
   b. coordination between sensors and actuators [4], as well as
   c. coordination between networks and contexts.
2. *Cyber-enabled homogeneity*—which refers to ubiquitous things with features developing towards an unity of the same nature driven in the cyber world—such as
   a. resource homogeneity,
   b. service homogeneity like mining-as-a-service (MaaS) [5], and
   c. functionality homogeneity.
3. *Social-inspired interaction*—which refers to human-oriented applications in both cyber and physical worlds—such as
   a. interaction between a real individual (in the physical world) and a cyber individual (in the cyber world) like personalized service recommendations and service discovery [6–9], as well as
   b. interaction among multiple cyber individuals like social relationships and interactions [10, 11].
4. *Brain-abstracted cooperation*—which refers to the Internet of (cognitive) thinking [12].

Consequently, these interconnections within the smart world have led to emerging research areas like:

- big data and smart computing [13–17],
- big data applications and services [18–22],
- cloud computing [23–27],
- data mining and analytics [28–34],

- data science [35–40],
- smart data [41], as well as
- social computing [42–45].

With these emerging research areas, in this paper, we focus on *big data computing and mining*, which aims to analyze and discover knowledge from rich sets of the aforementioned big data within the smart world.

We organize the remainder of this paper as follows. Starting from the next section, we discuss how frequent pattern mining interconnects with physical-cyber worlds. Then, in Sect. 3, we continue our discussion by adding interconnections of physical-cyber worlds with the thinking world via cognitive-based frequent pattern mining. Along this direction, in Sect. 4, we add interconnections of physical-cyber-thinking worlds with the social world via frequent social pattern mining. In Sect. 5, we discuss how these interconnections among different worlds can help in understanding and mining big epidemic data with an aim to combat diseases like coronavirus disease 2019 (COVID-19). Finally, we draw our conclusions in Sect. 6.

2 Interconnection Between the Physical World and the Cyber World via Frequent Pattern Mining

Among big data computing and mining tasks, we focus—in this paper—on frequent pattern mining, which aims to discover new knowledge in the form of collections of frequently co-occurring items (e.g., popular shopper market baskets of merchandise items, collated events, influential social groups). These frequent patterns serve as fundamental building blocks for many other big data computing and mining tasks in the smart world. For example,

- frequent patterns can be served as antecedents and consequents of interesting *association rules*, which reveal relationships between the frequent patterns on the antecedent and consequent parts of the association rules (e.g., reveal customer behavior).
- Along this direction, to extend further, frequent patterns can be served as antecedents and consequents of *associative classification* rules, which help predict the consequent labels based on frequent patterns on the antecedent parts of the classification rules (e.g., predict and make recommendations).

With advanced information and communication technology (ICT) in the *cyber world*, huge amounts of data are easily generated or gathered at a rapid rate from a broad range of rich data sources for numerous big data applications and services (e.g., in business, engineering, health, humanities natural sciences, social sciences) in the *physical world*. For instance, merchandise items that we placed in a shopper market basket for the in-store (physical) or virtual/online (cyber) purchase can be recorded in a transaction. Collections of frequently co-occurring items in these baskets can be discovered by frequent pattern mining algorithms in the cyber world. The mined frequent patterns then reveal customer purchase behaviour, which in turn helps in item inventory, promotions, and shelving.

2.1 Serial Algorithms

To find collections of frequently co-occurring items from the *physical world*, the Apriori algorithm [46] in the *cyber world* applies a breath-first approach to analyze the transactions and discover the corresponding frequent patterns in a level-wise bottom-up fashion. To elaborate, the algorithm first discovers frequent singletons from domain items (e.g., individual merchandise items in shopper market). It then keeps discovering frequent non-singletons, ranging from frequent pairs, to frequent triplets, then frequent quadruplets, and so on.

To expedite the frequent pattern mining process, the FP-growth algorithm [47] applies a depth-first approach to analyze the transactions and discover the corresponding frequent patterns. To elaborate, the algorithm first discovers frequent singletons domain items (e.g., individual merchandise items in shopper market). It then captures all transactions in a data set in a tree structure such that each tree path represents a transaction. Such a tree representation enables tree path sharing for common items "shared" (i.e., occurred) among transactions. The algorithm discovers non-singletons from this tree. Specifically, for each frequent singleton $\{x\}$, the algorithm keeps "growing" (i.e., extending) the frequent singleton to discover its related frequent pairs, triplets, quadruplets, and so on. In other words, it discovers all frequent non-singletons of different cardinality related to the frequent singleton $\{x\}$ before discovering its counterparts related to another frequent singleton $\{y\}$.

Both the Apriori and FP-growth algorithms in the cyber world discover frequent patterns using a *transaction-centric* paradigm, in which each transaction—for analyses and services in the physical world—captures the items in the transaction. Between them, Apriori discovers frequent patterns in a breath-first fashion, whereas FP-growth does so in a depth-first fashion.

As alternatives to these transaction-centric algorithms, *item-centric* algorithms [48] in the cyber world captures—for each pattern X—the IDs of all transaction IDs containing X. Then, the algorithms discover frequent patterns from the captured data. Among those well-known item-centric algorithms, for each pattern X, VIPER [49] uses a fixed-size bitmap vector to indicate IDs of transactions containing X. As such, the memory requirement for mining frequent patterns with VIPER is independent of the density of data. Frequent patterns of higher cardinality can be discovered by applying the bitwise AND operator to relevant sets of transaction IDs of lower cardinality. In contrast, for each pattern X, Eclat [50] captures a set of IDs of transactions containing X so that frequent patterns of higher cardinality can be discovered by intersecting relevant sets of transaction IDs of lower cardinality. This representation works well for sparse data set when the set of transaction IDs would be small. Similarly, dEclat [51] also captures a set of transaction IDs. More precisely, it captures the set difference in the set of transaction IDs containing X and those containing its extension $\mathrm{Y} = \mathrm{X} \cup \{x\}$ where X is grown by extending X with singleton $\{x\}$ to become Y.

To take advantage of these three individual algorithms, a serial hybrid vertical mining algorithm [52] switches from one frequent pattern algorithm to another during an iterative mining process. For instance, when the data set is relatively sparse in an iteration, the hybrid algorithm delays the switching from Eclat (which captures transaction IDs) to dEclat. On the other hand, when the data set becomes relatively dense in another iteration, the hybrid algorithm switches from Eclat to dEclat (which captures the set difference between transaction IDs of pattern X and its extension). Moreover, when the data set becomes denser in an iteration, the hybrid algorithm switches to VIPER (which capture the presence of transaction IDs in a fix-size bitmap). Hence, this hybrid algorithm in the cyber world discovers frequent patterns from data sets collected and generated in the physical world efficiently.

2.2 Parallel, Distributed, and High Performance Computing Algorithms

To further expedite the mining process of the aforementioned serial algorithms, parallel and distributed frequent pattern mining algorithms [53–57] were developed in the cyber world. Generally, they distribute data and/or patterns among parallel processors (i.e., workers) and return the discovered frequent patterns to the users through the master node in a cluster of distributed or parallel computers connected via networks such as local area networks (LANs). Along this direction, frequent pattern mining can also be conducted in a grid of computers.

In the recent decade, algorithms [58, 59] (e.g., SVT [60]) based on the MapReduce programming model [61] have become popular. As implied by its name, the MapReduce model uses the map and reduce functions. In particular, the master node partitions the input data set, and sends the partitioned data to the worker node in a *cloud computing environment* [23–27]. Then, each worker node applies the map function to obtain pairs of $\langle \text{key}, \text{value} \rangle$ [62]. Afterwards, the worker node applies the reduce function to the sorted $\langle \text{key}, \text{value} \rangle$ -pairs to obtain a reduced (e.g., aggregated such as summed) value for each key. In the context of frequent pattern mining, the map function outputs each occurrence of a pattern, and the reduce function computes the frequency of each pattern. Those patterns with sufficient frequency (i.e., reaches the user-specified threshold) are considered to be frequent. The map and reduce functions may be applied multiple times during the mining process. Here, it is interesting to observe the interaction of physical-cyber-thinking worlds in frequent pattern mining: The mining algorithms in the *cyber world* incorporates the user-specified threshold from the *thinking world* to discover frequent patterns from data generated and gathered in the *physical world*. For the aforementioned *cloud computing-based* frequent pattern mining, local data generated or collected by local Internet of Things (IoT) devices (e.g., sensors) [4, 63, 64] are usually sent to data cloud centers for mining.

2.3 Fog and Edge Computing Algorithms

With the technological advancements in the cyber world, the local data for analyses and services in the physical world can be transmitted to fog, which refers to local networking services located between the IoT devices and data cloud. This leads to *fog computing-based* frequent pattern mining [63], with which frequent patterns can be discovered in the fog (rather than the cloud).

Along this direction, when the local IoT devices have become more powerful, the mining can be performed on the local IoT devices. This leads to *edge computing-based* frequent pattern mining [65, 66], with which frequent patterns are discovered on the edge near the data source (rather than the fog or cloud).

3 Interconnection Between the Physical-Cyber Worlds and the Thinking World via Cognitive Mining

3.1 Frequent Pattern Mining-as-a-Service

Although the aforementioned algorithms in the cyber world are effective in mining frequent patterns, they were built for experts (e.g., data miners, data scientists) who have some degrees of data mining knowledge and expertise. However, there are situations where users are experts in their domains but not necessary expert data miners. To broaden the scope and availability of big data science—especially, frequent pattern mining—to these non-expert miners, frequent pattern mining-as-a-service (FPMaaS) [5] provides scalable service on cloud. Specifically,

- on the front end, it provides users (e.g., layman) in the *thinking world* with a friendly web interface for users to express their ways of cognitive thinking via user-specified queries; and
- on the back end, it incorporates the aforementioned algorithms in the *cyber world* to discover frequent patterns from data generated or collected for analyses and services in the *physical world*.

3.2 Constrained Frequent Pattern Mining via Crowdsourcing

When discovered frequent patterns from data for real-life big data applications and services in the *physical world*, the aforementioned algorithms in the *cyber world* have incorporated user cognitive thinking in the *thinking world* via the use of user-specified (frequency) threshold that determines whether a pattern is frequent. There are situations where users could express more from their cognitive thinking (than just the user-specified threshold)—via the use of user-specified constraints in addition to the (frequency) threshold. Examples include:

- Find the popular shopper market basket with merchandise items that worth more than $100 in total.
- Find the popular shopper market basket with the cheapest merchandise item of price at least $10.
- Find the popular shopper market basket with at most two promotional merchandise items.

A naïve approach is to first handle these user-specified constraints is discover all frequent patterns, then remove those not satisfying the user-specified constraints, and thus leaving and outputting those valid *constrained frequent patterns* (*CFP*) [67–70] —i.e., patterns that are frequent and satisfying the user-specified constraints). A better approach is to push constraints inside the mining process by exploiting properties of the user-specified (non-frequency) constraints.

As a third approach, Crowd-CFP-Miner [67] applies crowdsourcing and collaborative filtering to discover frequent patterns satisfying the user-specified constraints. Specifically, by applying the concept of crowd-voting,

- Crowd-CFP-Miner allows the vendors or service providers to "outsource" the frequent pattern mining task to only the targeted crowds of public users (e.g., based on the user profile and/or users' voluntary information), and
- Crowd-CFP-Miner also allows these targeted crowds to express their opinions—in terms of interests and preference (e.g., their interested items, their preferred items)—via online votes. Crowd-voting can be conducted by allowing the crowds to select their perceived popular or interesting items (or sets of items) from lists of (Boolean and/or categorical) choices.

By doing so, Crowd-CFP-Miner also broadens the scope and availability of big data science—especially, frequent pattern mining—to these non-expert miners. Both the concepts of FPMaaS and this Crowd-CFP-Miner algorithm in the *cyber world* incorporates users' cognition from the *thinking world* to discover frequent patterns from data generated or collected for analyses and services in the *physical world*.

4 Interconnection Between the Physical-Cyber-Thinking Worlds and the Social World via Social Network Analysis

Social media and social networks are rich sources of big data for many real-life applications. Consequently, although we have so far used the shopper market basket analysis as an illustrative example for frequent pattern mining in this paper, frequent pattern mining is also applicable for other analyses and services such as social network analysis. For example, mining algorithms [71–74] have been proposed in the cyber world to discover useful knowledge revealing social interactions among social individuals or groups in the social world, from the social data generated and collected in the physical world.

Align with the aforementioned concept of FPMaaS, some works were designed for non-expert miners to discover interesting patterns. For example, Leung et al. [75] applied big data analytics (from the cyber world) to social network data in the physical world to find out who cares about them on a social networking site (Facebook). Mai et al. [76] extended the idea to apply big data analytics to data from a social networking site (Twitter) for discover who is talking about their profession.

5 Discussion: Mining COVID-19 Data in a Smart World Environment

A smart world can be considered as a convergence of physical world, cyber world, social world, and thinking world. Interaction among these four worlds can be observed in various real-life applications for analyses, services and smart data such as

- urban analytics [17, 77], as well as
- disease, healthcare, and omic analytics [78–85].

For example, efficient and accurate algorithms could be designed and developed in the cyber world, with an aim to discover commonality and trends from an integration of

- epidemic and genomic data—collected in the physical world—for diseases like coronavirus disease 2019 (COVID-19, which caused pandemic in 2020);
- useful information from the social world (e.g., patients' physical and cyber social networks, their public profiles and public posts on social networking sites like Facebook and Twitter); and
- cognitive thinking of related personnel (e.g., healthcare providers, scientists like epidemiologists, public health officers) in the thinking world.

Such an interconnection of the cyber, physical, social, and thinking worlds would lead us into a better smart world.

6 Conclusions

A smart world is a convergence of physical world, cyber world, social world, and thinking world. In this paper, we presented and highlighted this convergence of physical-cyber-social-thinking worlds via the illustrative data science task of frequent pattern mining. Many algorithms developed in the cyber world discover frequent patterns from big data generated and collected from the physical world. Some of them incorporate user preference (e.g., users' votes) capturing user cognitive thinking from their thinking world via friendly interfaces. Moreover, these frequent pattern mining algorithms can be adapted for social network analysis, which reveals interactions among social individuals (and their groups) in the social world. This shows a tight connection among the four worlds—namely, the physical, cyber, social, and thinking worlds—within a smart world.

As ongoing and future work, we are exploring effective interaction among these four worlds in various real-life applications for analyses, services and smart data. Such an effective interaction among the cyber, physical, social, and thinking worlds—via big data computing and mining—would lead us into a better smart world environment.

Acknowledgements. This project is partially supported by (i) Natural Sciences and Engineering Research Council of Canada (NSERC) as well as (ii) University of Manitoba.

References

1. Ma, J., Yang, L.T., Apduhan, B.O., Huang, R., Barolli, L., Takizawa, M.: Towards a smart world and ubiquitous intelligence: a walkthrough from smart things to smart hyperspaces and UbicKids. Int. J. Pervasive Comput. Commun. **1**(1), 53–68 (2005). https://doi.org/10.1108/17427370580000113
2. Ning, H., Liu, H., Ma, J., Yang, L.T., Wan, Y., Ye, X., Huang, R.: From Internet to smart world. IEEE Access **3**, 1994–1999 (2015). https://doi.org/10.1109/ACCESS.2015.2493890
3. Ning, H., Liu, H.: Cyber-physical-social-thinking space based science and technology framework for the Internet of Things. Sci. China Inf. Sci. **58**(3), 1–19 (2015). https://doi.org/10.1007/s11432-014-5209-2
4. Leung, C.K., Braun, P., Cuzzocrea, A.: AI-based sensor information fusion for supporting deep supervised learning. Sensors **19**(6), 1345:1–1345:12 (2019). https://doi.org/10.3390/s19061345

5. Han, Z., Leung, C.K.: FIMaaS: scalable frequent itemset mining-as-a-service on cloud for non-expert miners. In: BigDAS 2015, pp. 84–91. ACM (2015). https://doi.org/10.1145/2837060.2837072
6. Guo, L., Yin, H., Wang, Q., Cui, B., Huang, Z., Cui, L.: Group recommendation with latent voting mechanism. In: IEEE ICDE 2020, pp. 121–132 (2020). https://doi.org/10.1109/ICDE48307.2020.00018
7. Jiang, F., Leung, C.K., Pazdor, A.G.M.: Web page recommendation based on bitwise frequent pattern mining. In: IEEE/WIC/ACM WI 2016, pp. 632–635. IEEE (2016). https://doi.org/10.1109/WI.2016.0111
8. Leung, C.K., Jiang, F., Souza, J.: Web page recommendation from sparse big web data. In: IEEE/WIC/ACM WI 2018, pp. 592–597. IEEE (2018). https://doi.org/10.1109/WI.2018.00-32
9. Leung, C.K., Kajal, A., Won, Y., Choi, J.M.C.: Big data analytics for personalized recommendation systems. In: IEEE DASC-PiCom-CBDCom-CyberSciTech 2019, pp. 1060–1065 (2019). https://doi.org/10.1109/DASC/PiCom/CBDCom/CyberSciTech.2019.00190
10. Fariha, A., Ahmed, C.F., Leung, C.K., Abdullah, S.M., Cao, L.: Mining frequent patterns from human interactions in meetings using directed acyclic graphs. In: PAKDD 2013, Part I. LNCS (LNAI), vol. 7818, pp. 38–49. Springer, Heidelberg (2013). https://doi.org/10.1007/978-3-642-37453-1_4
11. Leung, C.K., Tanbeer, S.K., Cameron, J.J.: Interactive discovery of influential friends from social networks. Soc. Netw. Anal. Min. **4**(1), 154:1–154:13 (2014). https://doi.org/10.1007/s13278-014-0154-z
12. Deng, D., Mai, J.J., Leung, C.K., Cuzzocrea, A.: Cognitive-based hybrid collaborative filtering with rating scaling on entropy to defend shilling influence. In: ICNCC 2019, pp. 176–185. ACM (2019). https://doi.org/10.1145/3375998.3376040
13. Audu, A.A., Cuzzocrea, A., Leung, C.K., MacLeod, K.A., Ohin, N.I., Pulgar-Vidal, N.C.: An intelligent predictive analytics system for transportation analytics on open data towards the development of a smart city. In: CISIS 2019. AISC, vol. 993, pp. 224–236. Springer, Cham (2019). https://doi.org/10.1007/978-3-030-22354-0_21
14. Kitagawa, H., Saad, W., Shim, K., Tang, J. (eds.): Proceedings of the IEEE BigComp 2019 (2019)
15. Lee, W., et al. (eds.): Proceedings of the IEEE BigComp 2020 (2020)
16. Leung, C.K.: Big data mining and computing in a smart world. In: IEEE UIC-ATC-ScalCom-CBDCom-IoP 2015, pp. ciii (2015). https://doi.org/10.1109/UIC-ATC-ScalCom-CBDCom-IoP.2015.341
17. Leung, C.K., Braun, P., Hoi, C.S.H., Souza, J., Cuzzocrea, A.: Urban analytics of big transportation data for supporting smart cities. In: DaWaK 2019. LNCS, vol. 11708, pp. 24–33. Springer, Cham (2019). https://doi.org/10.1007/978-3-030-27520-4_3
18. Lee, W., Leung, C.K. (eds.): Big Data Applications and Services 2017. AISC, vol. 770. Springer, Singapore (2017). https://doi.org/10.1007/978-981-13-0695-2
19. Lee, W., Leung, C.K. (eds.): Special Issue on Selected Papers from Smart Data 2018 and Big Data Service 2018. Sensors, vol. 19. MDPI (2019)
20. Leung, C.K.: Big data mining applications and services. In: BigDAS 2015, pp. 1–8. ACM (2015). https://doi.org/10.1145/2837060.2837076
21. Leung, C.K., Nasridinov, A. (eds.): Proceedings of the BigDAS 2015. ACM (2015)
22. Xu, L., Guo, T., Dou, W., Wang, W., Wei, J.: An experimental evaluation of garbage collectors on big data applications. PVLDB **12**(5), 570–583 (2019). https://doi.org/10.14778/3303753.3303762
23. Bellatreche, L., Leung, C.K., Xia, Y., Elbaz, D.: Advances in cloud and big data computing. Concurr. Comput.: Pract. Exp. **31**(2), e5053:1–e5053:3 (2019). https://doi.org/10.1002/cpe.5053

24. Hilprecht, B., Binnig, C., Röhm, U.: Learning a partitioning advisor for cloud databases. In: ACM SIGMOD 2020, pp. 143–157 (2020). https://doi.org/10.1145/3318464.3389704
25. Jiang, F., Leung, C.K.: A data analytic algorithm for managing, querying, and processing uncertain big data in cloud environments. Algorithms **8**(4), 1175–1194 (2015). https://doi.org/10.3390/a8041175
26. Jiang, F., Leung, C.K., Middleton, R., Pazdor, A.G.M.: Big social data mining in a cloud computing environment. In: ICCBB 2018, pp. 58–65. IEEE (2018). https://doi.org/10.1109/ICCBB.2018.8756461
27. Kobusinska, A., Leung, C.K., Hsu, C., Raghavendra, S., Chang, V.: Emerging trends, issues and challenges in Internet of Things, big data and cloud computing. Future Gener. Comput. Syst. (FGCS) **87**, 416–419 (2018). https://doi.org/10.1016/j.future.2018.05.021
28. Bleifuß, T., Bornemann, L., Johnson, T., Kalashnikov, D.V., Naumann, F., Srivastava, D.: Exploring change - a new dimension of data analytics. PVLDB **12**(2), 85–98 (2018). https://doi.org/10.14778/3282495.3282496
29. Cuzzocrea, A., Mumolo, E., Leung, C.K., Grasso, G.M.: An innovative deep-learning algorithm for supporting the approximate classification of workloads in big data environments. In: IDEAL 2019, Part II. LNCS, vol. 11872, pp. 225–237. Springer, Cham (2019). https://doi.org/10.1007/978-3-030-33617-2_24
30. Dewan, U., Ahmed, C.F., Leung, C.K., Rizvee, R.A., Deng, D., Souza, J.: An efficient approach for mining weighted frequent patterns with dynamic weights. In: ICDM 2019, pp. 13–27 (2019)
31. Lakshmanan, L.V.S., Leung, C.K., Ng, R.T.: The segment support map: scalable mining of frequent itemsets. ACM SIGKDD Explor. **2**(2), 21–27 (2000). https://doi.org/10.1145/380995.381005
32. Leung, C.K.: Uncertain frequent pattern mining. In: Frequent Pattern Mining, pp. 417–453. Springer, Cham (2014). https://doi.org/10.1007/978-3-319-07821-2_14
33. Leung, C.K., Mateo, M.A.F., Brajczuk, D.A.: A tree-based approach for frequent pattern mining from uncertain data. In: PAKDD 2008. LNCS (LNAI), vol. 5012, pp. 653–661. Springer, Heidelberg (2008). https://doi.org/10.1007/978-3-540-68125-0_61
34. Yoo, K., Leung, C.K., Nasridinov, A. (eds.): Special Issue on Big Data Analysis and Visualization. Applied Sciences, vol. 10. MDPI (2020)
35. Al-Dubai, A., et al. (eds.): Proceedings of the IEEE IUCC-DSCI-SmartCNS 2019 (2019)
36. Chen, J., Yang, LT. (eds.): Proceedings of the IEEE DSDIS-CPSCom-GreenCom-iThings 2015 (2015)
37. Dierckens, K.E., Harrison, A.B., Leung, C.K., Pind, A.V.: A data science and engineering solution for fast k-means clustering of big data. In: IEEE TrustCom-BigDataSE-ICESS 2017, pp. 925–932 (2017). https://doi.org/10.1109/TrustCom/BigDataSE/ICESS.2017.332
38. Leung, C.K., Zhang, Y.: An HSV-based visual analytic system for data science on music and beyond. IJACDT **8**(1), 68–83 (2019). https://doi.org/10.4018/IJACDT.2019010105
39. Parameswaran, A.: Enabling data science for the majority. PVLDB **12**(12), 2309–2322 (2019). https://doi.org/10.14778/3352063.3352148
40. Sarumi, O., Leung, C.K.: Scalable data science and machine learning algorithm for gene prediction. In: BigDAS 2019, pp. 118–126 (2019)
41. Leung, C.K., Zhang, Y., Hoi, C.S.H., Souza, J., Wodi, B.H.: Big data analysis and services: visualization of smart data to support healthcare analytics. In: IEEE iThings-GreenCom- CPSCom-SmartData 2019, pp. 1261–1268 (2019). https://doi.org/10.1109/iThings/GreenCom/CPSCom/SmartData.2019.00212
42. Jiang, F., Leung, C.K., Tanbeer, S.K.: Finding popular friends in social networks. In: CGC 2012, pp. 501–508. IEEE (2012). https://doi.org/10.1109/CGC.2012.99

43. Kim, J., Guo, T., Feng, K., Cong, G., Khan, A., Choudhury, F.M.: Densely connected user community and location cluster search in location-based social networks. In: ACM SIGMOD 2020, pp. 2199-2209. https://doi.org/10.1145/3318464.3380603
44. Leung, C.K.: Mathematical model for propagation of influence in a social network. In: Encyclopedia of Social Network Analysis and Mining, 2nd edn. pp. 1261–1269. Springer, New York (2018). https://doi.org/10.1007/978-1-4939-7131-2
45. Tanbeer, S.K., Leung, C.K., Cameron, J.J.: Interactive mining of strong friends from social networks and its applications in e-commerce. JOCEC **24**(2–3), 157–173 (2014). https://doi.org/10.1080/10919392.2014.896715
46. Agrawal, R., Srikant, R.: Fast algorithms for mining association rules in large databases. In: VLDB 1994, pp. 487–499. Morgan Kaufmann (1994)
47. Han, J., Pei, J., Yin, Y.: Mining frequent patterns without candidate generation. In: ACM SIGMOD 2000, pp. 1–12 (2000). https://doi.org/10.1145/335191.335372
48. Braun, P., Cuzzocrea, A., Leung, C.K., Pazdor, A.G.M., Souza, J.: Item-centric mining of frequent patterns from big uncertain data. Procedia Comput. Sci. **126**, 1875–1884 (2018). https://doi.org/10.1016/j.procs.2018.08.075
49. Shenoy, P., Bhalotia, J.R., Bawa, M., Shah, D.: Turbo-charging vertical mining of large databases. In: ACM SIGMOD 2000, pp. 22–33 (2000). https://doi.org/10.1145/335191.335376
50. Zaki, M.J.: Scalable algorithms for association mining. IEEE TKDE **12**(3), 372–390 (2000). https://doi.org/10.1109/69.846291
51. Zaki, M.J.: Fast vertical mining using diffsets. In: ACM KDD 2003, pp. 326–335 (2003). https://doi.org/10.1145/956750.956788
52. Leung, C.K.: Pattern mining for knowledge discovery. In: IDEAS 2019, pp. 34:1–34:5. ACM (2019). https://doi.org/10.1145/3331076.3331099
53. Cuzzocrea, A., Leung, C.K., Jiang, F., MacKinnon, R.K.: Complex mining from uncertain big data in distributed environments: problems, definitions and two effective and efficient algorithms. In: Big Data Management and Processing, pp. 297-332. Taylor & Francis (2017) https://doi.org/10.1201/9781315154008-15
54. Kumar, S., Mohbey, K.K.: A review on big data based parallel and distributed approaches of pattern mining. JKSU-CIS (2019). https://doi.org/10.1016/j.jksuci.2019.09.006
55. Leung, C.K., Jiang, F., Pazdor, A.G.M.: Bitwise parallel association rule mining for web page recommendation. In: IEEE/WIC/ACM WI 2017, pp. 662–669. ACM (2017). https://doi.org/10.1145/3106426.3106542
56. Quoc, P.H.V., Küng, J.: FPO tree and DP3 algorithm for distributed parallel frequent itemsets mining. ESWA **140**, 112874:1–112874:13 (2020). https://doi.org/10.1016/j.eswa.2019.112874
57. Zaki, M.J.: Parallel and distributed association mining: a survey. IEEE Concurr. **7**(4), 14–25 (1999). https://doi.org/10.1109/4434.806975
58. Braun, P., Cuzzocrea, A., Jiang, F., Leung, C.K., Pazdor, A.G.M.: MapReduce-based complex big data analytics over uncertain and imprecise social networks. In: DaWaK 2017. LNCS, vol. 10440, pp. 130–145. Springer, Cham (2017). https://doi.org/10.1007/978-3-319-64283-3_10
59. Luna, J.M., Padillo, F., Pechenizkiy, M., Ventura, S.: Apriori versions based on MapReduce for mining frequent patterns on big data. IEEE Trans. Cybern. **48**(10), 2851–2865 (2018). https://doi.org/10.1109/TCYB.2017.2751081
60. Leung, C.K., Zhang, H., Souza, J., Lee, W.: Scalable vertical mining for big data analytics of frequent itemsets. In: DEXA 2018, Part I. LNCS, vol. 11029, pp. 3–17. Springer, Cham (2018) https://doi.org/10.1007/978-3-319-98809-2_1
61. Dean, J., Ghemawat, S.: MapReduce: simplified data processing on large clusters. CACM **51**(1), 107–113 (2008). https://doi.org/10.1145/1327452.1327492

62. Hoi, C.S.H., Leung, C.K., Tran, K., Cuzzocrea, A., Bochicchio, M., Simonetti, M.: Supporting social information discovery from big uncertain social key-value data via graph-like metaphors. In: ICCC 2018. LNCS (LNISA), vol. 10971, pp. 102-116. Springer, Cham (2018). https://doi.org/10.1007/978-3-319-94307-7_8
63. Braun, P., Cuzzocrea, A., Leung, C.K., Pazdor, A.G.M., Souza, J., Tanbeer, S.K.: Pattern mining from big IoT data with fog computing: models, issues, and research perspectives. In: IEEE/ACM CCGrid 2019, pp. 854–891. IEEE (2019). https://doi.org/10.1109/CCGRID.2019.00075
64. Lee, D., Cho, J., Park, D.: Efficient partitioning of on-cloud remote executable code and on-chip software for complex-connected IoT. In: IEEE BigComp 2019, pp. 627–630 (2019). https://doi.org/10.1109/BIGCOMP.2019.8679228
65. Leung, C.K., Deng, D., Hoi, C.S.H., Lee, W.: Constrained big data mining in an edge computing environment. In: Big Data Applications and Services 2017. AISC, vol. 770, pp. 61–68. Springer, Singapore (2017). https://doi.org/10.1007/978-981-13-0695-2_8
66. Zhou, D., Ouyang, M., Kuang, Z., Li, Z., Zhou, J.P., Cheng, X.: Incremental association rule mining based on matrix compression for edge computing. IEEE Access **7**, 173044–173053 (2019). https://doi.org/10.1109/ACCESS.2019.2956823
67. Hoi, C.S.H., Khowaja, D., Leung, C.K.: Constrained frequent pattern mining from big data via crowdsourcing. In: Big Data Applications and Services 2017. AISC, vol. 770, pp. 69–79. Springer, Singapore (2017). https://doi.org/10.1007/978-981-13-0695-2_9
68. Leung, C.K.: Frequent itemset mining with constraints. In: Encyclopedia of Database Systems, 2nd edn. pp. 1531–1536. Springer, New York (2018) https://doi.org/10.1007/978-1-4614-8265-9_17
69. Nijssen, S., Zimmermann, A.: Constraint-based pattern mining. In: Frequent Pattern Mining, pp. 147–163. Springer, Cham (2014) https://doi.org/10.1007/978-3-319-07821-2_7
70. Sarumi, O.A., Leung, C.K.: Exploiting anti-monotonic constraints for mining palindromic motifs from big genomic data. In: IEEE BigData 2019, pp. 4864–4873 (2019). https://doi.org/10.1109/BigData47090.2019.9006397
71. Fan, C., Hao, H., Leung, C.K., Sun, L.Y., Tran, J.: Social network mining for recommendation of friends based on music interests. In: IEEE/ACM ASONAM 2018, pp. 833–840. IEEE (2018). https://doi.org/10.1109/ASONAM.2018.8508262
72. Leung, C.K., Cuzzocrea, A., Mai, J.J., Deng, D., Jiang, F.: Personalized DeepInf: enhanced social influence prediction with deep learning and transfer learning. In: IEEE BigData 2019, pp. 2871–2880 (2019). https://doi.org/10.1109/BigData47090.2019.9005969
73. Singh, S.P., Leung, C.K., Jiang, F., Cuzzocrea, A.: A theoretical approach to discover mutual friendships from social graph networks. In: iiWAS 2019, pp. 212–221. ACM (2019). https://doi.org/10.1145/3366030.3366077
74. Zarrinkalam, F., Fani, H., Bagheri, E.: Social user interest mining: methods and applications. In: ACM KDD 2019, pp. 3235–3236 (2019). https://doi.org/10.1145/3292500.3332279
75. Leung, C.K., Jiang, F., Poon, T.W., Crevier, P.: Big data analytics of social network data: who cares most about you on Facebook? In: Highlighting the Importance of Big Data Management and Analysis for Various Applications, pp. 1–15. Springer, Cham (2018) https://doi.org/10.1007/978-3-319-60255-4_1
76. Mai, M., Leung, C.K., Choi, J.M.C., Kwan, L.K.R.: Big data analytics of Twitter data and its application for physician assistants: who is talking about your profession in Twitter? In: Data Management and Analysis, pp. 17–32. Springer, Cham (2020). https://doi.org/10.1007/978-3-030-32587-9_2
77. Pan, Z., Liang, Y., Wang, W., Yu, Y., Zheng, Y., Zhang, J.: Urban traffic prediction from spatio-temporal data using deep meta learning. In: ACM KDD 2019, pp. 1720–1730 (2019). https://doi.org/10.1145/3292500.3330884

78. De Guia, J., Devaraj, M., Leung, C.K.: DeepGx: deep learning using gene expression for cancer classification. In: IEEE/ACM ASONAM 2019, pp. 913–920. ACM (2019). https://doi.org/10.1145/3341161.3343516
79. Han, P., Yang, P., Zhao, P., Shang, S., Liu, Y., Zhou, J., Gao, X., Kalnis, P.: GCN-MF: disease-gene association identification by graph convolutional networks and matrix factorization. In: ACM KDD 2019, pp. 705–713 (2019). https://doi.org/10.1145/3292500.3330912
80. Kulkarni, A., Sood, K., Kaul, S., Vasuja, R. (eds.): Big Data Analytics in Healthcare. Springer, Cham (2019). https://doi.org/10.1007/978-3-030-31672-3
81. Lee, P.: The unreasonable effectiveness, and difficulty, of data in healthcare. In: ACM KDD 2019, pp. 3–4 (2019). https://doi.org/10.1145/3292500.3330645
82. Pawliszak, T., Chua, M., Leung, C.K., Tremblay-Savard, O.: Operon-based approach for the inference of rRNA and tRNA evolutionary histories in bacteria. BMC Genom. **21**(Supplement 2), 252:1–252:14 (2020). https://doi.org/10.1186/s12864-020-6612-2
83. Sarumi, O.A., Leung, C.K., Adetunmbi, A.O.: Spark-based data analytics of sequence motifs in large omics data. Procedia Comput. Sci. **126**, 596–605 (2018). https://doi.org/10.1016/j.procs.2018.07.294
84. Souza, J., Leung, C.K., Cuzzocrea, A.: An innovative big data predictive analytics framework over hybrid big data sources with an application for disease analytics. In: AINA 2020. AISC, vol. 1151, pp. 669–680. Springer, Cham (2020). https://doi.org/10.1007/978-3-030-44041-1_59
85. Wu, P., Cheng, C., Kaddi, C.D., Venugopalan, J., Hoffman, R., Wang, M.D.: Omic and electronic health record big data analytics for precision medicine. IEEE Trans. Biomed. Eng. **64**(2), 263–273 (2017). https://doi.org/10.1109/TBME.2016.2573285

Data Science for Big Data Applications and Services: Data Lake Management, Data Analytics and Visualization

Carson K. Leung(✉)

University of Manitoba, Winnipeg, MB, Canada
kleung@cs.umanitoba.ca

Abstract. Huge amounts of useful data are easily generated and gathered currently at a rapid rate from a broad range of rich data sources in numerous applications and services in the real world. Data science applies database techniques, scientific and engineering methods, mathematical and statistical models, data mining algorithms, and/or machine learning tools to manage data, extract the useful information and discover the new knowledge from these big data. This explains why data science for big data applications and services has become a fundamental technology in providing novel solutions in various areas in business, engineering, health, humanities, natural sciences, social sciences, etc. (e.g., healthcare, manufacturing, social life). Usually, data science focuses on big data management, analytics and visualization. Once big data are managed (i.e., captured, curated, managed and processed), big data are analyzed with an aim to discover interesting knowledge and information, which is usually presented in text or table form. Consistent with a proverb that "a picture is worth a thousand words", big data visualization as well as visual analytics helps to reveal and explain the discovered interesting knowledge and information. In this paper, we present (a) big data management with focus on information fusion and the data lake; (b) big data analytics and mining, with focus on frequent patterns; as well as (c) big data visualization with focus on a few visual analytic systems for visualizing big data and mined frequent patterns. For illustration, we discuss these three aspects of data science on coronavirus disease 2019 (COVID-19) data. This highlights some important aspects of data science for big data analyses, services, and smart data.

Keywords: Visual analytics · Data visualization · Outcome visualization · Big data · Data science · Data management · Data lake · Big data applications · Big data services · COVID-19

1 Introduction and Related Works

Rapid progress in computing and information technology enables us to collect or generate huge amounts of useful data (e.g., business, financial, healthcare, manufacturing, shopper market basket, and social data) at different degrees of veracities (e.g., precise data, uncertain and imprecise data) at a rapid rate (e.g., streaming data) from a broad range of

W. Lee et al. (Eds.): *Big Data Analyses, Services, and Smart Data*, AISC 899, pp. 28–44, 2021.
https://doi.org/10.1007/978-981-15-8731-3_3

rich data sources in numerous real-life business, engineering, health, scientific and social applications and services [1–5]. The big data are usually characterized by well-known V's (e.g., 3 V's, ..., 7 V's, ..., 42 V's, 51 V's) [6] such as:

- volume, which characterizes the *data quantity* (e.g., huge amounts of data [7]);
- value, which characterizes the *data usefulness* (e.g., valuable information and useful knowledge that can be mined or discovered from the big data [8–12]);
- veracity, which characterizes the *data quality* (e.g., precise data such as shopper market basket data; uncertain and imprecise data [13–19] such as medical diagnoses, sensor readings, voice picked up by speech recognition devices, writings recognized by optical character readers; privacy-preserving data [20–24]);
- velocity, which characterizes the *data generation or collection speed* (e.g., dynamic streams of big data [25–28], sequences [29, 30]);
- variety, which characterizes the differences in *data contents, formats* or *types* (e.g., graphs [31–33], key-value pairs [34]);
- validity, which characterizes the *data interpretation* and knowledge interpretation [35–38]; and
- visibility, which characterizes the *data visualization* and knowledge visualization [35–38].

Data science applies database techniques, scientific and engineering methods, mathematical and statistical models, and/or data mining and machine learning (e.g., deep learning) algorithms to

- manage the big data;
- extract useful information and discover new knowledge from the big data; and
- interpret and visualize the extracted information and the discovered knowledge.

Hence, data science for big data applications and services has become a fundamental technology in providing novel solutions for many areas in the fields of business, engineering, health, humanities, natural sciences, social sciences, etc. Usually, data science focuses on

- big data management (e.g., information fusion, data lake),
- big data analytics (e.g., data mining), and
- big data visualization (e.g., visual analytics, frequent pattern visualization).

In the remainder of this paper, we describe these three key aspects of data science. In particular, we describe big data management, with focus on information fusion and the data lake, in Sect. 2. Then, we describe big data analytics and mining in Sect. 3. Afterwards, we present big data visualization, with focus on visual analytics of mined frequent patterns, in Sect. 4. In Sect. 5, we discuss the roles of these key aspects of data science in a big data application and service of outbreak analytics. In particular, we discuss how manage, analyze and visualize big epidemic data and their knowledge, with an aim to combat diseases like coronavirus disease 2019 (COVID-19). Finally, we draw our conclusions in Sect. 6.

2 Big Data Management: Information Fusion and Data Lake

Recall from Sect. 1, big data can of different types, contents, and/or formats collected or generated for different data sources, *big data management* aims to capture, curate, manage and process (e.g., clean, filter) data. This leads to the concepts of *data integration* and *information fusion* [39, 40], which combines big data with information from a rich collection of homogeneous or heterogeneous data sources to obtain accurate, consistent and useful knowledge.

The combined data can be put in a *data lake* [41–44], which is repository of data stored in its natural/raw format. A data lake generally stores

- *structured data* from relational databases (organized in rows and columns);
- *semi-structured data* in formats like comma-separated values (CSV), extensible markup languages (XML), JavaScript object notation (JSON), and logs;
- *unstructured data* (e.g., emails, documents, images, digital audio, and video files).

These big data need to be properly managed. Otherwise, a data lake may turn into a *data swamp* (referring to a deteriorated or unmanaged data lake that becomes inaccessible to the intended users or of limited/no value). Some of these properly managed data from the data lake can be stored in a structured form in

- a *data warehouse* (or an *enterprise data warehouse*), or
- simply a *data mart*

for further analyses and/or reporting.

As an illustrative example, consider an international retail company with branches in different parts of the world. Merchandise items sold in some part of the world (e.g., France) may be described in French text with dimensions & weights measured in metric system, whereas items sold in another part of the world (e.g., USA) may be captured as images with dimensions & weights measured in the Imperial system. These textual and image data can be stored in a data lake, which can then be standardized (e.g., with equivalent English descriptions, dimensions & weights measured in units—like cm and kg—in the metric system). Their processed data can be captured in a data warehouse for further data analytics.

3 Big Data Analytics and Mining

Once big data are stored and properly managed in a data lake, *big data analytics and mining* then apply computational (e.g., data mining, machine learning) algorithms, mathematical methods, and/or statistical models to analyze the data and to extract useful information and discover new knowledge from the data. Over the past 2–3 decades, numerous data mining algorithms have been designed and developed for finding interesting patterns like frequent patterns, association rules, clusters, classifiers, predictions, and/or anomalies.

Recall from Sect. 1, amounts of big data are usually huge. This leads to *scalable* big data analytic and mining algorithms and techniques [45–47] —including those use parallel and distributed computing, high-performance computing (e.g., cloud, cluster, grid computing), as well as edge and fog computing [10, 48–51]. Moreover, as big data are generated or gathered at a rapid rate, many *incremental big data analytic* and *big data stream mining* algorithms [25–28] have been designed and implemented. Furthermore, big data can also be (a) precise or (b) uncertain and imprecise. This leads to *uncertain* data mining [13–19].

Continue with our illustrative example on an international retail company. Scalable frequent pattern mining algorithms help discover sets of popular (i.e., frequently purchased) merchandise items from shopper market baskets. Extensions of these algorithms (e.g., stream mining and/or uncertain data mining) help data analysts to mine frequent patterns from big data streams and/or big uncertain data. The discovered knowledge from these patterns would have economic and social impacts in our daily life and in various applications and services.

4 Big Data Visualization: Visual Analytics

Once the big data are managed and analyzed, interesting information and knowledge (e.g., frequent patterns in our illustrative example) is discovered from the big data. Usually, the discovered knowledge is represented in a text or tabular form (e.g., a lengthy list of frequent patterns), which may not be comprehendible to many users. Consistent with a proverb that "a picture is worth a thousand words", big data visualization and *visual analytics* helps to reveal and explain the discovered interesting knowledge and information. So, we present a few visual analytic systems—in this section—for big data and result visualization—especially, for visualization and visual analytics of frequent patterns from big data.

Many tools and systems [36, 52] have been designed and developed over the past 2–3 decades to visualize data and/or the analyzed result (e.g., discovered knowledge):

- data (e.g., VisDB [52]), and
- new knowledge discovered from big data (e.g., clusters, decision trees) [53–55].

Among them, some tools and systems were built to visualize patterns revealing knowledge related to frequently occurring sets of items (i.e., itemsets) such as popular shopper market baskets of merchandise products or collocated events. For instance, Munzner et al. [56] designed PowerSetViewer (PSV), which is a visualization system for visualizing powersets of items. In PSV, users are guaranteed with visibility of a certain frequent patterns (i.e., frequent itemsets). In terms of representation, PSV vertically arranges the collections of patterns in ascending order of cardinality so that (a) shorter patterns (e.g., frequent itemsets of cardinality 1) appear on the top of the screen whereas (b) longer patterns (e.g., frequent itemsets of the highest cardinality K) appear on the bottom. Among the patterns of the same cardinality for each cardinality, they are sorted lexicographically to facilitate a quick lookup. Moreover, PSV uses colour to show the frequency of a pattern. However, despite the existence of connections among related patterns (e.g., superset-subset relationships), PSV does not explicitly display these connections.

As another instance, Wong et al. [57] designed visualization tools to visualize association rules, as well as sequential patterns, related to topics that appear in documents. Similar to parallel coordinates, the visualization tools display (a) keywords on the parallel vertical axes (i.e., y-axes) and (b) their sequential (e.g., temporal) index on the x-axis. In other words, instead of visualizing frequent patterns, the visualization tools display association rules and sequential patterns (mined from documents).

As a third instance, Yang [58] also designed a system for visualizing association rules. He did so by representing association rules in a two-dimensional space that consists of parallel vertical axes. To adapt his system for visualizing frequent patterns, all items in the domain can be sorted in descending frequency order before distributing them evenly along each vertical axis. With this arrangement, a curve extending from one vertical axis to another, spanning over k related parallel vertical axes, represents a frequent pattern consisting of the corresponding k items (i.e., a k-itemset). The thickness of such a curve reveals the frequency (i.e., occurrence counts) of the frequent k-itemset represented by the curve. However, it can be challenging to compare the thickness of several curves, and thus the frequencies of several frequent patterns.

4.1 Big Data Visualization via Polylines or Orthogonal Wires

To allow users to easily compare frequencies of several frequent patterns, FIsViz [59] displays each frequent k-itemset in the form a polyline—instead of a curve—that connects k nodes in a two-dimensional space. Crucial information about any frequent k-itemset is captured by its associated (x, y)-coordinates in this two-dimensional space. Specifically, FIsViz lists all items in the domain on the x-axis, and lists the frequency of all frequent patterns on the y-axis. With this arrangement, the x-positions of all nodes in the polyline reveal all the items (from the domain) within the corresponding frequent pattern Z, and the y-position of the rightmost node of a polyline for Z explicitly reveals the frequency of Z. In addition to explicit indication of frequencies of frequent patterns, FIsVIz also explicitly indicates association among frequent patterns such as the prefix-extension relationships when users traverse along the polylines representing these frequent patterns. As shown in Fig. 1(a), prefixes—as well as their frequencies—of Z can be visualized by traversing leftward along the polyline for Z. Similarly, extensions—as well as their frequencies—of Z can be visualized by traversing rightward along the polyline for Z.

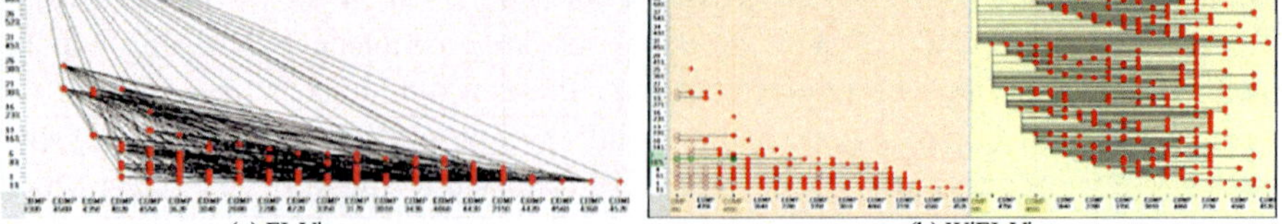

(a) FIsViz (b) WiFIsViz

Fig. 1. Big data visualization via (a) polylines in FIsViz [59] and (b) orthogonal wires in WiFIsViz [60].

Observing that polylines in FIsViz may not be easily distinguishable from one another due to their potential bending and crossing-over for frequent patterns mined from big data, WiFIsViz [60] displays each frequent *k*-itemset in a horizontal line that connects *k* nodes in a two-dimensional space in a wiring-type diagram (i.e., orthogonal graph). Specifically, as shown in Fig. 1(b), WiFIsViz deploys two half-screens to display frequent patterns. Here, the left half-screen provides users with frequencies frequent patterns, whereas the right half-screen provides users with prefix-extension relationships among frequent patterns in the form of a trie. Like FIsViz, WiFIsViz also lists all items in the domain on the *x*-axes of both half-screens. However, it only lists frequency values of all frequent patterns on the *y*-axis of the left half-screen, in which the *y*-position of the horizontal line for a frequent pattern Z explicitly reveals the frequency of Z. It links frequent patterns between the two half-screens in the sense that, when a user examines a frequent pattern on either half-screen, the corresponding pattern are highlighted—in green in Fig. 1(b)—on another half-screen. For instance, when WiFIsViz reveals the frequency of Z on the left half-screen, it also highlights the corresponding Z in the trie to reveal prefixes of Z (via leftward traversals on the trie of Z) and extensions of Z (via rightward traversals on the trie of Z) on the right half-screen.

4.2 Hierarchical Big Data Visualization

As an alternative to frequent pattern visualization via polylines or orthogonal wires, PyramidViz [61] visualizes frequent patterns in a hierarchical way. Specifically, it emphasizes on visualizing the association among frequent patterns such as the prefix-extension relationships. In particular, PyramidViz displays frequent patterns in a hierarchical layout—namely, a building block layout—in such a way that

- short patterns (e.g., frequent itemsets of cardinality 1) are put on the bottom of the pyramid,
- related longer patterns (e.g., frequent itemsets of higher cardinality which are extensions of their related prefixes) are put above them, and
- the longest related patterns (e.g., frequent itemsets of the highest cardinality K) are put on the top of the pyramid.

In other words, the building blocks for the prefixes (and extensions) of a frequent pattern Z are put below (and above) the building block that represents Z. Consequently, PyramidViz explicitly shows the association among frequent patterns such as the prefix-extension relationships. In addition, as shown in Fig. 2(a), PyramidViz—which provides users with a side-view of the pyramid—clearly labels items within each of these frequent patterns, and these labels are linked in the form of a trie connecting through the center of the associated building blocks. PyramidViz represents the frequency range of Z by the colour of the corresponding building block.

In contrast to PyramidViz (which visualizes frequent patterns with a *side-view* of the pyramid), FpMapViz [62] visualizes frequent patterns with a *top-view*. Specifically, FpMapViz represents frequent patterns as collections of rectangular-shape objects such as squares. As shown in Fig. 2(b), FpMapViz put frequent patterns in a hierarchical layout so that

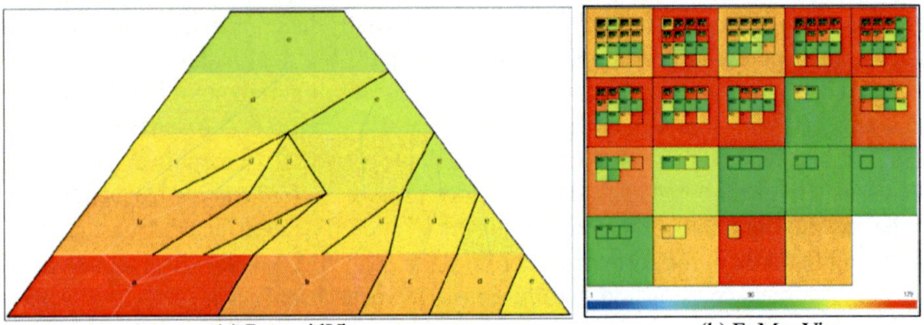

(a) PyramidViz (b) FpMapViz

Fig. 2. Hierarchical big data visualization: (a) PyramidViz [61] and (b) FpMapViz [62]

- squares representing short patterns (e.g., frequent itemsets of cardinality 1) are placed on the background of the screen, and
- squares representing the related longer patterns (e.g., frequent itemsets of higher cardinality which are extensions of their related prefixes) are placed on top (i.e., on the foreground).

As a side-benefit, FpMapViz shows the cardinality of a frequent pattern Z by the size of the square corresponding to Z. FpMapViz—as similar to PyramidViz—also represents the frequency range of Z by the colour (of the corresponding square).

4.3 Orientation-Free Big Data Visualization

The aforementioned visualization systems work well, especially for a single-user environment (in which the data and/or the mined results—say, in the form of frequent patterns—can be viewed individually) or multi-user environment (in which the data and/or the mined results can be projected in a upright position on the wall or screen). However, in today's collaborative environment, collaborators may sit around a table-top displays (on which information is shared) for in-person collaborative discussions. Note that information viewed from different orientations may be lead to different perceptions. Observed from a study [63], when a chart is not oriented properly, user perception (e.g., legibility or readability) of the chart decreases. As illustrated in Fig. 3, person A has the best view due to the perfect orientation with respect to person A. In contrast, both persons B and C have the side view of the chart, but person D views the chart upside down.

Fortunately, as an orientation-free visualizer, RadialViz [64, 65] improves the user perception. Specifically, although RadialViz also visualizes frequent patterns in a hierarchical fashion, it does so with a radial layout. Specifically,

- sectors representing short patterns (e.g., frequent itemsets of cardinality 1) are placed on the background of the screen, and
- sectors representing the related longer patterns (e.g., frequent itemsets of higher cardinality which are extensions of their related prefixes) are placed on top (i.e., on the foreground)

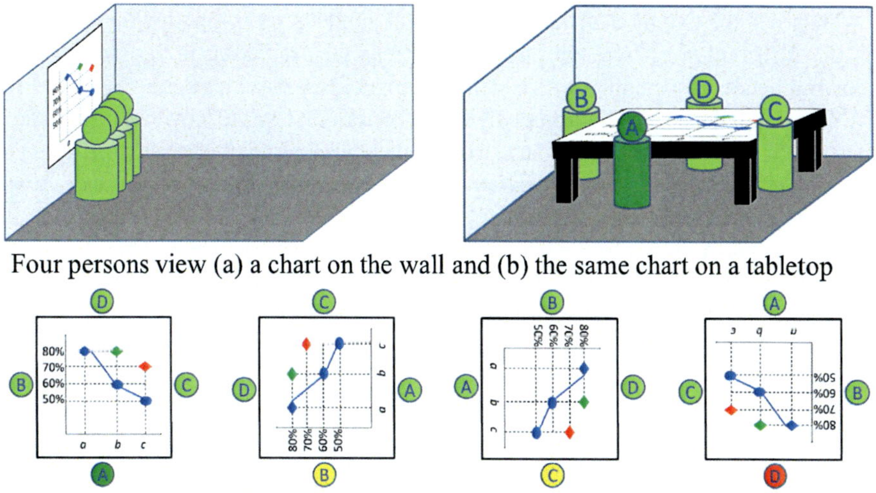

Four persons view (a) a chart on the wall and (b) the same chart on a tabletop

(c) A's view (d) B's view (e) C's view (f) D's view

Fig. 3. Illustrative example of the importance of having an orientation-free representation

Unlike both PyramidViz and FpMapViz (which represent the frequency range of a frequent pattern by its colour), RadialViz represents the cardinality of the frequent pattern by its colour. To ease of visualizing frequent patterns, as shown in Fig. 4(a), RadialViz provides users with two visual modes for showing the same collections of frequent patterns:

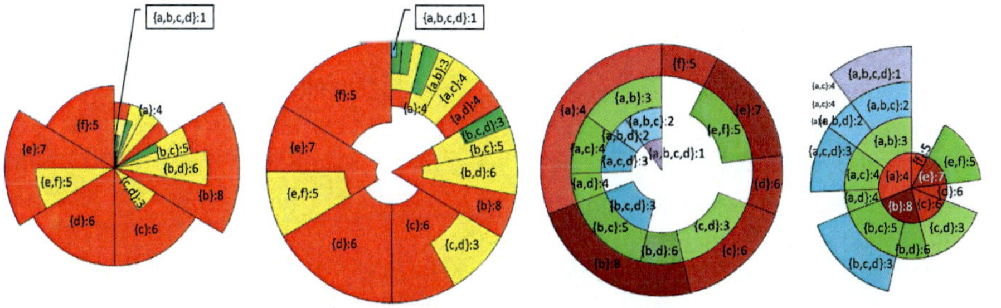

(a) RadialViz ("radiating out" & "radiating in" versions) (b) HSVis (c) HSLviz

Fig. 4. Orientation-free big data visualization: (a) RadialViz [64, 65], (b) HSVis [66], (c) HSLviz [67]

- "radiating out" mode, in which RadialViz represents the frequency of a frequent pattern by its radius (from the center); and
- "radiating in" mode, in which RadialViz first determines the radius by taking the maximum among the frequency of all frequent patterns, and then represents the frequency of a frequent pattern by the direct distance (from circumference toward the center) of the sector representing Z. Here, the radius is determined by maximum frequency among all frequent patterns.

Between the two modes, the "radiating out" mode is good for showing an overview on the distribution of frequent patterns, whereas the "radiating in" mode is good for showing details (especially, for frequent patterns of low cardinalities).

Observed that a single extremely frequent pattern that was discovered from a skewed data set (having an uneven distribution of frequencies among the frequent patterns) may lead to incomprehensive visualization of the frequent patterns. For instance, a spike with a long radius representing an extremely frequent pattern can lead to relative short radii for sectors representing other patterns, and thus not easy to be visible. Fortunately, both HSVis [66] (which was designed to visualize new knowledge acquired from music analytics such as popular collections of lyrics) and HSLviz [67] (which was designed to visualize new knowledge acquired from healthcare analytics such as common diseases with their correlated symptoms and habits) can be adapted to frequent pattern visualization and to solve this problem. They do so by avoiding the use of radius in the radial representation to show the frequency of frequent patterns. Instead, they use different properties of colour in a colour model to show the frequency and cardinality of frequent patterns.

More specifically, HSVis uses the hue-saturation-value (HSV) colour model, which is a subtractive colour model (in which ink or paint can be mixed together to produce a black colour in an edge case). Generally, the HSV colour model represents its three key colours properties—namely, hue, saturation and value—by a *single* (inverted) hexacone. Specifically:

- Hue possesses the *colour appearance* with respect to visual sensation. It ranges between 0° and 360° including 0° to represents colours such as red (when hue is around 0°), yellowish green (when hue is around 90°), cyan (when hue is around 180°) and violet (when hue is around 270°).
- Saturation possesses the *colour purity*. To indicate the hue intensity, saturation reveals the colourfulness of a stimulus with respect to its own brightness. It ranges from 0 to 1 inclusive: Purer colour appears at a higher saturation level (e.g., approaching or at 1), and more washed-out (e.g., whiter) appears at a lower saturation level (e.g., approaching or at 0).
- Value possesses the *colour brightness*. It also ranges from 0 to 1 inclusive: Darker (i.e., blacker) colour appears at a lower value (e.g., approaching or at 0). Since the HSV colour model is commonly represented by a single inverted hexacone, the colour is shaded to black when the value is at 0 (regardless of the ranges of hue or saturation).

When compared with the commonly used red-green-blue (RGB) colour model (in which red, green and blue light can be additively superimposed in different ways in producing a wide colour range), the HSV colour model has been observed to be more closely aligned with the human perception of the colour-making attributes in human vision. Hence, in the HSV colour model, HSVis visualizes frequent patterns with

- hue for representing the frequency of frequent patterns, and
- the combined saturation-value for representing the cardinality of frequent patterns.

As shown in Fig. 4(b), HSVis displays frequent patterns with a "radiating in" mode, in which HSVis first determines the number of rings by taking the maximum cardinality among all frequent patterns, places the frequent patterns of lower cardinalities on the outer rings, and places their extensions ring-by-ring inward in ascending order of cardinalities.

In contrast, HSLviz [67] uses the hue-saturation-lightness (HSL) colour model, which is an additive colour model like the RGB colour model (in which light sources can be additively superimposed to give a white colour in an edge case). Generally, the HSL colour model represents its three key colours properties—namely, hue, saturation and lightness—by *double* hexacones. Specifically,

- Lightness possesses the *perceived brightness* of a colour. It ranges from 0 to 1 inclusive: Lighter (i.e., whiter) colour appears at a higher lightness level (e.g., approaching or at 1), and darker (i.e., blacker) colour appears at a lower lightness level (e.g., approaching or at 0). Since the HSL colour model is commonly represented by double hexacones, the colour is (a) tinted to white when the lightness is at 1 and (b) shaded to black when the lightness is at 0 (regardless of the ranges of hue or saturation).
- To a further extent, with double hexacones, saturation in this HSL colour model is different from its counterpart in the aforementioned HSV colour model. Specifically, in the HSL colour model, purer colour appears when the saturation level is approaching or at 0.5, colour is more tint (e.g., the whiter) when increasing the saturation level from 0.5 (e.g., closer to or at 1), and the colour is more shade (e.g., the blacker) when decreasing the saturation level from 0.5 (e.g., closer to or at 0).

Like the commonly used RGB colour model, the HSL colour model is also supported by many computer programs. Hence, based on this HSL colour model, HSLviz has a benefit of avoiding extra conversion from one colour model to another (e.g., from HSV to either RGB or HSL), and thus saving the execution time. Hence, in the HSL colour model, HSLviz visualizes frequent patterns with

- hue for representing the frequency, and
- combined saturation-lightness for representing the cardinality of frequent patterns.

As shown in Fig. 4(c), HSLviz displays frequent patterns with a "radiating out" mode, in which HSLviz places the frequent patterns of lower cardinalities on the inner rings, and places their extensions ring-by-ring outward in ascending order of cardinalities.

4.4 Summary of Comparisons Among Visualizers

So far, we have described polyline-/wire-based big data visualizers (e.g., FIViz, WiFIsViz), hierarchical big data visualizers (e.g., PyramidViz, FpMapViz), and orientation-free big data visualizers (e.g., RadialViz, HSVis, HSLviz). We compare and summarize key characteristics of these big data visualizers in Table 1.

Table 1. Key characteristics of big data visualizers

	Cardinality	Frequency	Extension
FIsViz [59]	#nodes in a polyline	*y*-position of the rightmost node of a polyline	Traverse along the polylines
WiFIsViz [60]	#nodes in a horizontal line	*y*-position of the horizontal line on the left half-screen	Traverse along the trie on the right half-screen
PyramidViz [61]	Level at the pyramid from the bottom (for 1-itemsets)	Colour of the pyramid block	Above the pyramid block for the prefix
FpMapViz [62]	Size of the square	Colour of the square	On top of the square for the prefix
RadialViz [64, 65]	Colour of the sector	Radius of the sector	On top of the sector for the prefix
HSVis [66]	Hue	Saturation-value	On the inner ring of the prefix
HSLviz [67]	Hue	Saturation-lightness	On the outer ring of the prefix

5 Discussion: Data Science on COVID-19 Data

The three aforementioned key aspects have played an important role in data science for many real-life applications and services [68, 69] such as

- music and movie analytics [1, 66, 70–72];
- sports and game analytics [35, 73–77];
- learning analytics [59, 60, 78];
- transport analytics [79–81]; as well as
- healthcare, disease, and omic analytics [67, 82–86].

For example, a data science approach could be helpful in combating diseases like Zika fever (which caused epidemic in South America during 2015–2016) or coronavirus disease 2019 (COVID-19, which caused pandemic in 2020). To elaborate, heterogeneous data related to COVID-19 (e.g., epidemiological/public health data on the numbers of cases/deaths, diagnosis from medical images, protein and genomic structures of the coronavirus) could be integrated, stored and managed in a data lake. These data could then be analyzed and mined by big data analytics—say, via healthcare, disease, and omic analytics—for discovering useful knowledge (e.g., commonality, trends). Finally, to facilitate comprehensiveness of the discovered knowledge, the aforementioned visualizers could be used. This data science approach enables scientists to get better understanding of the data and the disease, and thus helps decision makes to take appropriate actions for control and prevention of the disease.

6 Conclusions

Rapid progress in computing and information technology enables us to collect or generate huge amounts of useful data (e.g., financial, healthcare, manufacturing, shopper market basket, and social data) at different degrees of veracities (e.g., precise data, uncertain and imprecise data) at a rapid rate (e.g., streaming data) from a broad range of rich data sources in numerous applications and services. In this paper, we first described the *big data management* tasks of data integration and information fusion, in which integrated data are put in a *data lake*. Some properly managed data from the data lake are then stored in a data warehouse or a data mart for further analyses and/or reporting.

With huge amounts of big data—which may be of different degrees of veracities—generated and collected at a rapid rate, we then described *big data analytics* with emphasis on data mining. In particular, we focused on scalable data mining (e.g., frequent pattern mining) algorithms for discovering useful information and new knowledge (e.g., frequent patterns) from big data streams and/or big uncertain data stored and managed in a data lake.

Knowing the importance of the validity and visibility of big data, we described *big data visualization*. It supports the interpretation and visualization of big data, as well as of the knowledge discovered by big data analytics. The seven presented big data visualizers show and visualize information like cardinality, frequency and extensions of frequent patterns differently.

As ongoing and future work, we are exploring effective big data management (e.g., data lake management), analytics and visualization techniques to further enhance the data science support to different real-life big data applications and services. Such an exploration would lead us into a better (e.g., safer and healthier) environment.

Acknowledgements. This project is partially supported by (i) Natural Sciences and Engineering Research Council of Canada (NSERC) as well as (ii) University of Manitoba.

References

1. Jang, S., Lkhagvadorj, B., Nasridinov, A.: Preference-aware music recommendation using song lyrics. In: Lee, W., Leung, C.K. (eds.) BIGDAS 2017. AISC, vol. 770, pp. 183–195. Springer, Singapore (2019). https://doi.org/10.1007/978-981-13-0695-2_18
2. Jiang, F., Leung, C.K., Tanbeer, S.K.: Finding popular friends in social networks. In: CGC 2012, pp. 501–508. IEEE (2012). https://doi.org/10.1109/CGC.2012.99
3. Leung, C.K.-S., Tanbeer, S.K., Cameron, J.J.: Interactive discovery of influential friends from social networks. Soc. Netw. Anal. Min. **4**(1), 1–13 (2014). https://doi.org/10.1007/s13278-014-0154-z
4. Ryu, G.-A., Lee, J.-W., Jeong, J.-S., Kim, M., Yoo, K.-H.: Real-time smart safe-return-home service based on big data analytics. In: Lee, W., Leung, C.K. (eds.) BIGDAS 2017. AISC, vol. 770, pp. 197–209. Springer, Singapore (2019). https://doi.org/10.1007/978-981-13-0695-2_19
5. Tanbeer, S.K., Leung, C.K., Cameron, J.J.: Interactive mining of strong friends from social networks and its applications in e-commerce. JOCEC **24**(2–3), 157–173 (2014). https://doi.org/10.1080/10919392.2014.896715

6. Khan, N., Naim, A., Hussain, M.R., Naveed, Q.N., Ahmad, N., Qamar, S.: The 51 V's of big data: survey, technologies, characteristics, opportunities, issues and challenges. In: COINS 2019, pp. 19–24. ACM (2019). https://doi.org/10.1145/3312614.3312623
7. Shin, W., Baek, N.: Design and implementation of a sunshine duration calculation system with massively parallel processing. In: Lee, W., Leung, C.K. (eds.) BIGDAS 2017. AISC, vol. 770, pp. 91–97. Springer, Singapore (2019). https://doi.org/10.1007/978-981-13-0695-2_11
8. Hoi, C.S.H., Khowaja, D., Leung, C.K.: Constrained frequent pattern mining from big data via crowdsourcing. In: Lee, W., Leung, C.K. (eds.) BIGDAS 2017. AISC, vol. 770, pp. 69–79. Springer, Singapore (2019). https://doi.org/10.1007/978-981-13-0695-2_9
9. Leung, C.K.: Frequent itemset mining with constraints. In: Liu, L, Özsu, M.T. (eds.) Encyclopedia of Database Systems, 2nd edn. pp. 1531–1536. Springer, New York (2018). https://doi.org/10.1007/978-1-4614-8265-9_17
10. Leung, C.K., Deng, D., Hoi, C.S.H., Lee, W.: Constrained big data mining in an edge computing environment. In: Lee, W., Leung, C.K. (eds.) BIGDAS 2017. AISC, vol. 770, pp. 61–68. Springer, Singapore (2019). https://doi.org/10.1007/978-981-13-0695-2_8
11. Li, H., Lee, J., Mi, H., Yin, M.: Finding good subtrees for constraint optimization problems using frequent pattern mining. In: AAAI 2020, pp. 1577–1584 (2020)
12. Wang, C., Zheng, X.: Application of improved time series Apriori algorithm by frequent itemsets in association rule data mining based on temporal constraint. Evol. Intell. **13**(1), 39–49 (2019). https://doi.org/10.1007/s12065-019-00234-5
13. Braun, P., Cuzzocrea, A., Leung, C.K., Pazdor, A.G.M., Souza, J.: Item-centric mining of frequent patterns from big uncertain data. Procedia Comput. Sci. **126**, 1875–1884 (2018). https://doi.org/10.1016/j.procs.2018.08.075
14. Leung, C.K.-S.: Uncertain frequent pattern mining. In: Aggarwal, C.C., Han, J. (eds.) Frequent Pattern Mining, pp. 339–367. Springer, Cham (2014). https://doi.org/10.1007/978-3-319-07821-2_14
15. Leung, C.K., Hoi, C.S.H., Pazdor, A.G.M., Wodi, B.H., Cuzzocrea, A.: Privacy-preserving frequent pattern mining from big uncertain data. In: IEEE BigData 2018, pp. 5101–5110 (2018). https://doi.org/10.1109/BigData.2018.8622260
16. Leung, C.K.-S., Mateo, M.A.F., Brajczuk, D.A.: A tree-based approach for frequent pattern mining from uncertain data. In: Washio, T., Suzuki, E., Ting, K.M., Inokuchi, A. (eds.) PAKDD 2008. LNCS (LNAI), vol. 5012, pp. 653–661. Springer, Heidelberg (2008). https://doi.org/10.1007/978-3-540-68125-0_61
17. Lin, J.C.-W., Li, T., Pirouz, M., Zhang, J., Fournier-Viger, P.: High average-utility sequential pattern mining based on uncertain databases. KAIS **62**(3), 1199–1228 (2019). https://doi.org/10.1007/s10115-019-01385-8
18. Ovi, J.A., Ahmed, C.F., Leung, C.K., Pazdor, A.G.M.: Mining weighted frequent patterns from uncertain data streams. In: Lee, S., Ismail, R., Choo, H. (eds.) IMCOM 2019. AISC, vol. 935, pp. 917–936. Springer, Cham (2019). https://doi.org/10.1007/978-3-030-19063-7_72
19. Rahman, M.M., Ahmed, C.F., Leung, C.K.: Mining weighted frequent sequences in uncertain databases. Inf. Sci. **479**, 76–100 (2019). https://doi.org/10.1016/j.ins.2018.11.026
20. Ma, C., Wang, B., Jooste, K., Zhang, Z., Ping, Y.: Practical privacy-preserving frequent itemset mining on supermarket transactions. IEEE Syst. J. **14**(2), 1992–2002 (2020). https://doi.org/10.1109/JSYST.2019.2922281
21. Qiu, S., Wang, B., Li, M., Liu, J., Shi, Y.: Toward practical privacy-preserving frequent itemset mining on encrypted cloud data. IEEE TCC **8**(1), 312–323 (2020). https://doi.org/10.1109/TCC.2017.2739146

22. Telikani, A., Gandomi, A.H., Shahbahrami, A., Dehkordi, M.N.: Privacy-preserving in association rule mining using an improved discrete binary artificial bee colony. ESWA **144**, 113097:1–113097:19 (2020). https://doi.org/10.1016/j.eswa.2019.113097
23. Teo, S.G., Cao, J., Lee, V.C.S.: DAG: a general model for privacy-preserving data mining. IEEE TKDE **32**(1), 40–53 (2020). https://doi.org/10.1109/TKDE.2018.2880743
24. Wodi, B.H., Leung, C.K., Cuzzocrea, A., Sourav, S.: Fast privacy-preserving keyword search on encrypted outsourced data. In: IEEE BigData 2019, pp. 6266–6275 (2019). https://doi.org/10.1109/BigData47090.2019.9046058
25. Cuzzocrea, A., Jiang, F., Leung, C.K., Liu, D., Peddle, A., Tanbeer, S.K.: Mining popular patterns: a novel mining problem and its application to static transactional databases and dynamic data streams. In: Hameurlain, A., Küng, J., Wagner, R., Cuzzocrea, A., Dayal, U. (eds.) Transactions on Large-Scale Data- and Knowledge-Centered Systems XXI. LNCS, vol. 9260, pp. 115–139. Springer, Heidelberg (2015). https://doi.org/10.1007/978-3-662-47804-2_6
26. Ishita, S.Z., Ahmed, C.F., Leung, C.K., Hoi, C.H.S.: Mining regular high utility sequential patterns in static and dynamic databases. In: Lee, S., Ismail, R., Choo, H. (eds.) IMCOM 2019. AISC, vol. 935, pp. 897–916. Springer, Cham (2019). https://doi.org/10.1007/978-3-030-19063-7_71
27. Leung, C.K., Khan, Q.I.: DSTree: a tree structure for the mining of frequent sets from data streams. In: IEEE ICDM 2006, pp. 928–932 (2006). https://doi.org/10.1109/ICDM.2006.62
28. Nguyen, T.T., Weidlich, M., Zheng, B., Yin, H., Nguyen, Q.V.H., Stantic, B.: From anomaly detection to rumour detection using data streams of social platforms. PVLDB **12**(9), 1016–1029 (2019). https://doi.org/10.14778/3329772.3329778
29. Lee, K.Y., Suh, Y.-K.: Efficient mining of time interval-based association rules. In: Lee, W., Leung, C.K. (eds.) BIGDAS 2017. AISC, vol. 770, pp. 121–125. Springer, Singapore (2019). https://doi.org/10.1007/978-981-13-0695-2_13
30. Rizvee, R.A., Shahin, M.S.H., Ahmed, C.F., Leung, C.K., Deng, D., Mai, J.J.: Sliding window based weighted periodic pattern mining over time series data. In: ICDM 2019, pp. 118–132 (2019)
31. Fariha, A., Ahmed, C.F., Leung, C.K.-S., Abdullah, S.M., Cao, L.: Mining frequent patterns from human interactions in meetings using directed acyclic graphs. In: Pei, J., Tseng, V.S., Cao, L., Motoda, H., Xu, G. (eds.) PAKDD 2013, Part I. LNCS (LNAI), vol. 7818, pp. 38–49. Springer, Heidelberg (2013). https://doi.org/10.1007/978-3-642-37453-1_4
32. Islam, M.A., Ahmed, C.F., Leung, C.K., Hoi, C.S.H.: WFSM-MaxPWS: an efficient approach for mining weighted frequent subgraphs from edge-weighted graph databases. In: Phung, D., Tseng, V.S., Webb, G.I., Ho, B., Ganji, M., Rashidi, L. (eds.) PAKDD 2018, Part III. LNCS (LNAI), vol. 10939, pp. 664–676. Springer, Cham (2018). https://doi.org/10.1007/978-3-319-93040-4_52
33. Song, J.J., Kang, I., Lee, W., Kim, J., Lee, J.: Discussions on subgraph ranking for keyworded search. In: IEEE Cybermatics 2018, pp. 935–936 (2018). https://doi.org/10.1109/Cybermatics_2018.2018.00179
34. Hoi, C.S.H., Leung, C.K., Tran, K., Cuzzocrea, A., Bochicchio, M., Simonetti, M.: Supporting social information discovery from big uncertain social key-value data via graph-like metaphors. In: Xiao, J., Mao, Z.-H., Suzumura, T., Zhang, L.-J. (eds.) ICCC 2018. LNCS, vol. 10971, pp. 102–116. Springer, Cham (2018). https://doi.org/10.1007/978-3-319-94307-7_8
35. Braun, P., Cuzzocrea, A., Keding, T.D., Leung, C.K., Pazdor, A.G.M., Sayson, D.: Game data mining: clustering and visualization of online game data in cyber-physical worlds. Procedia Comput. Sci. **112**, 2259–2268 (2017). https://doi.org/10.1016/j.procs.2017.08.141
36. Jentner, W., Keim, D.A.: Visualization and visual analytic techniques for patterns. In: Fournier-Viger, P., Lin, J.C.-W., Nkambou, R., Vo, B., Tseng, V.S. (eds.) High-Utility Pattern

Mining. SBD, vol. 51, pp. 303–337. Springer, Cham (2019). https://doi.org/10.1007/978-3-030-04921-8_12

37. Kovalerchuk, B.: Interpretable knowledge discovery reinforced by visual methods. In: ACM KDD 2019, pp. 3219–3220 (2019). https://doi.org/10.1145/3292500.3332278
38. Leung, C.K.: Data and visual analytics for emerging databases. In: Lee, W., Choi, W., Jung, S., Song, M. (eds.) Proceedings of the 7th International Conference on Emerging Databases. LNEE, vol. 461, pp. 203–213. Springer, Singapore (2018). https://doi.org/10.1007/978-981-10-6520-0_21
39. Choi, S., Cho, S.: Sensor information fusion by integrated AI to control public emotion in a cyber-physical environment. Sensors **18**, 3767:1–3767:21 (2018). https://doi.org/10.3390/s18113767
40. Lenzerini, M.: Data integration: a theoretical perspective. In: ACM SIGMOD-SIGACT-SIGART PODS 2002, pp. 233–246 (2002). https://doi.org/10.1145/543613.543644
41. Bogatu, A., Fernandes, A.A.A., Paton, N.W., Konstantinou, N.: Dataset discovery in data lakes. In: IEEE ICDE 2020, pp. 709–720 (2020). https://doi.org/10.1109/ICDE48307.2020.00067
42. Giudice, P.L., Musarella, L., Sofo, G., Ursino, D.: An approach to extracting complex knowledge patterns among concepts belonging to structured, semi-structured and unstructured sources in a data lake. Inf. Sci. **478**, 606–626 (2019). https://doi.org/10.1016/j.ins.2018.11.052
43. Nargesian, F., Zhu, E., Miller, R.J., Pu, K.Q., Arocena, P.C.: Data lake management: challenges and opportunities. PVLDB **12**(12), 1986–1989 (2019). https://doi.org/10.14778/3352063.3352116
44. Zhang, Y., Ives, Z.G.: Finding related tables in data lakes for interactive data science. In: ACM SIGMOD 2020, pp. 1951–1966 (2020). https://doi.org/10.1145/3318464.3389726
45. Hubail, M.A., Alsuliman, A., Blow, M., Carey, M.J., Lychagin, D., Maxon, I., Westmann, T.: Couchbase analytics: NoETL for scalable NoSQL data analysis. PVLDB **12**(12), 2275–2286 (2019). https://doi.org/10.14778/3352063.3352143
46. Lakshmanan, L.V.S., Leung, C.K., Ng, R.T.: The segment support map: scalable mining of frequent itemsets. ACM SIGKDD Explorations **2**(2), 21–27 (2000). https://doi.org/10.1145/380995.381005
47. Li, Y., Xu, W.: PrivPy: general and scalable privacy-preserving data mining. In: ACM KDD 2019, pp. 1299–1307 (2019). https://doi.org/10.1145/3292500.3330920
48. Bonomi, F., Milito, R., Zhu, J., Addepalli, S.: Fog computing and its role in the Internet of Things. In: MCC 2012, pp. 13–16. ACM (2012). https://doi.org/10.1145/2342509.2342513
49. Braun, P., Cuzzocrea, A., Leung, C.K., Pazdor, A.G.M., Souza, J., Tanbeer, S.K.: Pattern mining from big IoT data with fog computing: models, issues, and research perspectives. In: IEEE/ACM CCGrid 2019, pp. 854–891. IEEE (2019). https://doi.org/10.1109/CCGRID.2019.00075
50. Shi, W., Cao, J., Zhang, Q., Li, Y., Xu, L.: Edge computing: vision and challenges. IEEE Internet Things J. **3**(5), 637–646 (2016). https://doi.org/10.1109/JIOT.2016.2579198
51. Yousefpour, A., Fung, C., Nguyen, T., Kadiyala, K., Jalalid, F., Niakanlahiji, A., Kong, J., Jue, J.P.: All one needs to know about fog computing and related edge computing paradigms: a complete survey. J. Syst. Archit. **98**, 289–330 (2019). https://doi.org/10.1016/j.sysarc.2019.02.009
52. Keim, D.A., Kriegel, H.: Visualization techniques for mining large databases: a comparison. IEEE TKDE **8**(6), 923–938 (1996). https://doi.org/10.1109/69.553159
53. Ankerst, M., Elsen, C., Ester, M., Kriegel, H.: Visual classification: an interactive approach to decision tree construction. In: ACM KDD 1999, pp. 392–396 (1999). https://doi.org/10.1145/312129.312298

54. Hassan, M.R., Ramamohanarao, K., Karmakar, C., Hossain, M.M., Bailey, J.: A novel scalable multi-class ROC for effective visualization and computation. In: Zaki, M.J., Yu, J.X., Ravindran, B., Pudi, V. (eds.) PAKDD 2010, Part I. LNCS (LNAI), vol. 6118, pp. 107–120. Springer, Heidelberg (2010). https://doi.org/10.1007/978-3-642-13657-3_14
55. Kovalerchuk, B.: Visual Knowledge Discovery and Machine Learning. Springer, Cham (2018). https://doi.org/10.1007/978-3-319-73040-0
56. Munzner, T., Kong, Q., Ng, R.T., Lee, J., Klawe, J., Radulovic, D., Leung, C.K.: Visual mining of power sets with large alphabets. Technical report, UBC CS TR-2005-25 (2005)
57. Wong, P.C., Cowley, W., Foote, H., Jurrus, E., Thomas, J.: Visualising sequential patterns for text mining. In: IEEE InfoVis 2000, pp. 105–111 (2000). https://doi.org/10.1109/INFVIS.2000.885097
58. Yang, L.: Pruning and visualising generalized association rules in parallel coordinates. IEEE TKDE **17**(1), 60–70 (2005). https://doi.org/10.1109/TKDE.2005.14
59. Leung, C.K., Irani, P.P., Carmichael, C.L.: FIsViz: a frequent itemset visualizer. In: Washio, T., Suzuki, E., Ting, K.M., Inokuchi, A. (eds.) PAKDD 2008. LNCS (LNAI), vol. 5012, pp. 644-652. Springer, Heidelberg (2008). https://doi.org/10.1007/978-3-540-68125-0_60
60. Leung, C.K., Irani, P.P., Carmichael, C.L.: WiFIsViz: effective visualisation of frequent itemsets. In: IEEE ICDM 2008, pp. 875–880 (2008). https://doi.org/10.1109/ICDM.2008.93
61. Leung, C.K., Kononov, V.V., Pazdor, A.G.M.: PyramidViz: visual analytics and big data visualization of frequent patterns. In: IEEE DASC-PICom-DataCom-CyberSciTech 2016, pp. 913–916 (2016). https://doi.org/10.1109/DASC-PICom-DataCom-CyberSciTec.2016.158
62. Leung, C.K., Jiang, F., Irani, P.P.: FpMapViz: a space-filling visualization for frequent patterns. In: IEEE ICDM 2011 Workshops, pp. 804–811 (2011) https://doi.org/10.1109/ICDMW.2011.86
63. Alallah, F., Jin, D., Irani, P.: OA-graphs: orientation agnostic graphs for improving the legibility of charts on horizontal displays. In: ACM ITS 2010, pp. 211–220 (2010). https://doi.org/10.1145/1936652.1936692
64. Leung, C.K.-S., Jiang, F.: RadialViz: an orientation-free frequent pattern visualizer. In: Tan, P.-N., Chawla, S., Ho, C.K., Bailey, J. (eds.) PAKDD 2012, Part II. LNCS (LNAI), vol. 7302, pp. 322–334. Springer, Heidelberg (2012). https://doi.org/10.1007/978-3-642-30220-6_27
65. Dubois, P.M.J., Han, Z., Jiang, F., Leung, C.K.: An interactive circular visual analytic tool for visualization of web data. In: IEEE/WIC/ACM WI 2016, pp. 709–712. IEEE (2016). https://doi.org/10.1109/WI.2016.0127
66. Leung, C.K., Zhang, Y.: An HSV-based visual analytic system for data science on music and beyond. Int. J. Art Cult. Des. Technol. (IJACDT) **8**(1), 68–83 (2019). https://doi.org/10.4018/IJACDT.2019010105
67. Leung, C.K., Zhang, Y., Hoi, C.S.H., Souza, J., Wodi, B.H.: Big data analysis and services: visualization of smart data to support healthcare analytics. In: IEEE iThings-GreenCom-CPSCom-SmartData.2019, pp. 1261–1268 (2019). https://doi.org/10.1109/iThings/GreenCom/CPSCom/SmartData.2019.00212
68. Leung, C.K.: Big data analytics and mining for knowledge discovery. In: Encyclopedia of Organizational Knowledge, Administration, and Technology. IGI Global (2021). https://doi.org/10.4018/978-1-7998-3473-1
69. Meng, F., Younas, M., Sugumaran, V. (eds.): Proceedings of the IEEE BigDataService 2019 (2019)
70. Haughton, D., McLaughlin, M., Mentzer, K., Zhang, C.: Movie Analytics. Springer, Cham (2015) https://doi.org/10.1007/978-3-319-09426-7
71. Leung, C.K., Eckhardt, L.B., Sainbhi, A.S., Tran, C.T.K., Wen, Q., Lee, W.: A flexible query answering system for movie analytics. In: Cuzzocrea, A., Greco, S., Larsen, H.L., Saccà, D.,

Andreasen, T., Christiansen, H. (eds.) FQAS 2019. LNCS (LNAI), vol. 11529, pp. 250–261. Springer, Cham (2019). https://doi.org/10.1007/978-3-030-27629-4_24
72. Meredith, D. (ed.): Computational Music Analysis. Springer, Cham (2015). https://doi.org/10.1007/978-3-319-25931-4
73. Brown, J.A., Cuzzocrea, A., Kresta, M., Kristjanson, K.D.L., Leung, C.K., Tebinka, T.W.: A machine learning system for supporting advanced knowledge discovery from chess game data. In: IEEE ICMLA 2017, pp. 649–654 (2017). https://doi.org/10.1109/ICMLA.2017.00-87
74. Leung, C.K., Joseph, K.W.: Sports data mining: predicting results for the college football games. Procedia Comput. Sci. **35**, 710–719 (2014). https://doi.org/10.1016/j.procs.2014.08.153
75. Leung, C.K., Kanke, F., Cuzzocrea, A.: Data analytics on the board game Go for the discovery of interesting sequences of moves in joseki. Procedia Comput. Sci. **126**, 831–840 (2018). https://doi.org/10.1016/j.procs.2018.08.017
76. Morgulev, E., Azar, O.H., Lidor, R.: Sports analytics and the big-data era. Int. J. Data Sci. Anal. **5**(4), 213–222 (2018). https://doi.org/10.1007/s41060-017-0093-7
77. Seif El-Nasr, M., Drachen, A., Canossa, A. (eds.): Game Analytics. Springer, London (2013). https://doi.org/10.1007/978-1-4471-4769-5
78. El Atia, S., Ipperciel, D., Zaïane, O.R. (eds.): Data Mining and Learning Analytics. Wiley (2016) https://doi.org/10.1002/9781118998205
79. Antoniou, C., Dimitriou, L., Pereira, F. (eds.): Mobility Patterns, Big Data and Transport Analytics. Elsevier (2019). https://doi.org/10.1016/C2016-0-03572-6
80. Leung, C.K., Braun, P., Hoi, C.S.H., Souza, J., Cuzzocrea, A.: Urban analytics of big transportation data for supporting smart cities. In: Ordonez, C., Song, I.-Y., Anderst-Kotsis, G., Tjoa, A.M., Khalil, I. (eds.) DaWaK 2019. LNCS, vol. 11708, pp. 24–33. Springer, Cham (2019). https://doi.org/10.1007/978-3-030-27520-4_3
81. Ukkusuri, S., Yang, C. (eds.): Transportation Analytics in the Era of Big Data. Springer, Cham (2019). https://doi.org/10.1007/978-3-319-75862-6
82. Arivaradarajan, P., Misra, G. (eds.): Omics Approaches, Technologies And Applications. Springer, Singapore (2018). https://doi.org/10.1007/978-981-13-2925-8
83. El Morr, C., Ali-Hassan, H. (eds.): Analytics in Healthcare. Springer, Cham (2019) https://doi.org/10.1007/978-3-030-04506-7
84. Pawliszak, T., Chua, M., Leung, C.K., Tremblay-Savard, O.: Operon-based approach for the inference of rRNA and tRNA evolutionary histories in bacteria. BMC Genom. **21**(Supplement 2), 252:1–252:14 (2020). https://doi.org/10.1186/s12864-020-6612-2
85. Reddy, C.K., Aggarwal, C.C. (eds.): Healthcare Data Analytics. Chapman and Hall/CRC (2015). https://doi.org/10.1201/b18588
86. Souza, J., Leung, C.K., Cuzzocrea, A.: An innovative big data predictive analytics framework over hybrid big data sources with an application for disease analytics. In: Barolli, L., Amato, F., Moscato, F., Enokido, T., Takizawa, M. (eds.) AINA 2020. AISC, vol. 1151, pp. 669–680. Springer, Cham (2020). https://doi.org/10.1007/978-3-030-44041-1_59

Detection of Editing Bursts and Extraction of Significant Keyphrases from Wikipedia Edit History

Zihang Chen and Mizuho Iwaihara(✉)

Graduate School of Information, Production and Systems, Waseda University, Kitakyushu, Japan
chenzihang@fuji.waseda.jp, iwaihara@waseda.jp

Abstract. In an online collaboration system such as Wikipedia, edit history is stored as revisions. Topics of articles or categories grow and fade over time, and evolutionary information is retained in its edit history. We consider that a great amount of information that is related to real life events is embedded in such edit history of documents. This paper focuses on a particular temporal text mining task: effectively extracting keyphrases from burst periods in the edit history of Wikipedia articles or category. We first combine the ARIMA model with a decay function to find typical edit burst periods, then perform keyphrase extraction on burst periods to reveal topics of bursts. However, keyphrase extraction methods, such as TextRank, do not consider temporal trends in text stream. In this paper, we propose TextRank_nfidf which reflects temporal trends into phrase node weights, by computing smoothed difference of editing frequency between revisions. We confirm that detected bursts and keyphrases are matching well with events along the timeline.

Keywords: Edit history · Burst detection · Keywords extraction · TextRank

1 Introduction

Wikipedia, one of the most popular websites, is a free, multilingual Internet encyclopedia written collaboratively by volunteers around the world [15], almost anyone can create or edit articles in Wikipedia, a large number of users actively participate in editing articles of Wikipedia, making Wikipedia store ever growing amounts of information. Everyone can edit or create articles in Wikipedia under its rules. For each article in Wikipedia, edit activities are recorded and stored as revisions in Wikipedia edit history. A revision serves as a snapshot of an article at specific time point, containing information such as timestamps, article title, contributor, text content, edit comment, etc.

Real-world events attracting massive public attention, such as political scandals, and natural disasters often trigger creation of Wikipedia articles, and succeeding developments and findings of the events produce revisions lasting for days or months. Wikipedia editors select significant topics from new media and other sources, where duplicates are removed. So we can expect that revisions in Wikipedia edit history hold more compact and summarized descriptions than web information sources. Revisions contain

W. Lee et al. (Eds.): *Big Data Analyses, Services, and Smart Data*, AISC 899, pp. 45–65, 2021.
https://doi.org/10.1007/978-981-15-8731-3_4

keyphrases that can explain bursting edits. Discovering such bursting keyphrases is beneficial for grasping the developments of the target event. However, one Wikipedia category that holds articles related to the target event often contain hundreds of articles, and one article can have hundreds of revisions as the article grows over time. Therefore, it is necessary to develop an efficient method for finding editing bursts and keyphrases related to the bursts, from a large collection of revisions.

A volume of work show that large amounts of information can be extracted by detection and analysis of bursts among the revisions of articles in Wikipedia [5, 7]. In this paper, we focus on extracting keyphrases closely related to bursts from a given collection of articles. First, we discuss detecting bursts to find the time period that has a significantly high frequency of edits. Then we show a method to extract keyphrases from revisions in each burst period. In this part, we propose TextRank_nfidf which considers both edge weights and node weights of a word graph, where a node weight reflects the past edit activity level of the corresponding phrase. Our algorithm shows superior performance over TextRank. We contrast quality of extracted phrases with Google Trends, where the results show that keyphrases we extracted are well representing real-life events in burst periods. Google Trends is a reflection of search heat of Internet users, where a trend reflects real-time search popularity of the target phrase. On the other hand, edit activities on Wikipedia articles reflect continuing developments of events, combined with continuous refinement of articles. Thus, keyphrases extracted from article revisions reflect factual significance evaluated by editors, where development may continue even after people's attention fades.

We also show that our method is applicable to find bursting keyphrases from a relatively small article set, such as a Wikipedia category. This is an advantage such that we can precisely specify an article set to be monitored, and bursting patterns on phrases can be compared over different categories. We further show that edit activities are migrating between article categories.

A previous version of this paper appeared in [4]. This paper is extended mainly on discussions on bursty keyphrases over multiple article sets and migration of edit activities on them, in Sect. 5.

The rest of this paper is organized as follows. We introduce related work about keyword extraction and burst detection in Sect. 2. In Sect. 3 we describe the general framework and details of our proposing methods. Experimental evaluation results are shown in Sect. 4. In Sect. 5, we discuss migration of edit activities between article categories. We address a conclusion and discuss future directions in Sect. 6.

2 Related Work

Al Tamime, Reham, et al. [1] focus on detecting burstiness in Wikipedia articles during disease outbreaks. They apply the ARIMA model [3] on the time series of the number of revisions and unique editors. Their model shows high accuracy in detecting editing bursts. Qin Yu et al. [12] employ an exponential weighted average method on the ARIMA model for anomaly detection in traffic of wireless sensor networks. It is demonstrated that their short-step exponential weighted average method can raise the correct anomaly detection rate.

Liu, Ruoran, et al. [9] are utilizing temporal information and topic information to mine evolution phases of hot events. They identify occurrence, development, climax, decline, and ending phases during the lifespan of hot events.

Inspired by PageRank [11] algorithm, Mihalcea and Tarau [10] proposed TextRank, which extracts keywords by co-occurrence relationship in words. TextRank constructs an edge weighted word graph where ranking scores are evaluated based on the PageRank algorithm. Bellaachia and Al-Dhelaan [2] added node weights to TextRank, which improves the precision in extracting keywords in document sets. But these keyword extraction methods do not consider the temporal information of edits on documents. In our work, we redefine the node weights and assign smoothed edit activity levels as node weights of words.

Y. Liu, et al. [16] proposed a keyphrase extraction method based on phrase frequency over history (PFH) and survival rate (SR). They construct term frequency time series by Kleinberg's burst detection algorithm and apply k-means with dynamic time warping (DTW) distance to cluster phrases having similar burst patterns. DTW can match similar patterns even with certain time delay, but DTW is time consuming, therefore although their method achieved good clustering results, it can only be applied to several articles, cannot handle hundreds of articles under a category of Wikipedia. Also, the optimal setting of k for k-means is not easy to determine.

A number of researches apply word segmentation [6] and chunking to keyphrase extraction [8], which are based on statistics of part-of-speech. Noun word groups are meaningful and easy for human to understand. In our work, we use part-of-speech (POS) Tagger [14] to divide each sentence of revisions into chunks, where chunks containing noun POS tags are considered as candidate nodes of our TextRank_nfidf.

3 Proposed Method

Our objective is to find keyphrases in an editing burst period of articles in a given Wikipedia category of interest. Therefore, given edit histories of a given set of Wikipedia articles under a certain category or topic, first, we find editing burst periods through the ARIMA_EWA model, and then we need to extract related keyphrases through our improved TextRank algorithm for each burst period. Figure 1 shows the framework of our proposed method.

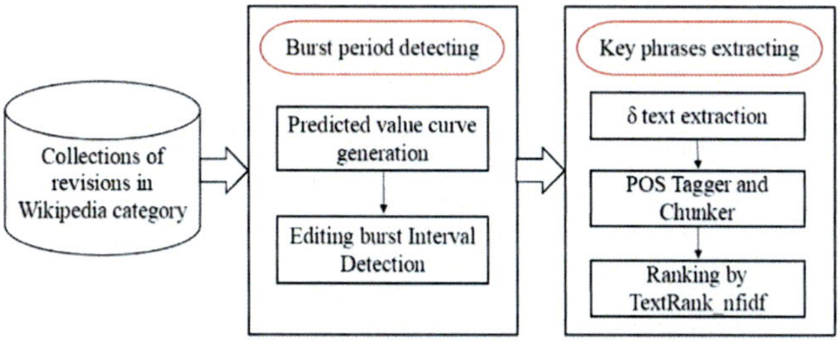

Fig. 1. Framework of our proposed method

3.1 Burst Period Detection

The ARIMA model was developed by Box and Jenkins in 1970 [3]. Non-seasonal ARIMA models are denoted by ARIMA(p, d, q), where non-negative integer parameters p, d, and q, are such that p is the order of the autoregressive model, d is the degree of differencing, and q is the order of the moving average model [12]. For time series $\{X_{t,}\ t = 1, 2, \ldots, n\}$, ARIMA($p$, d, q) can be expressed as:

$$(1 - \sum_{i=1}^{p} \phi_i B^i)(1 - L)^d\ X_t = (1 + \sum_{i=1}^{q} \theta_i B^i)\varepsilon_t \tag{1}$$

Here B denotes backward shift operator, Φ_1, .., Φ_p are the parameters of the autoregressive part of the model, θ_1 ,.., θ_q are the parameters of the moving average part, and ε_t is an error term. The error terms ε_t are generally assumed to be independent. By training ARIMA models with historical revision count data, we can detect spikes in the temporal sequence of revisions. We use a fixed size sliding window to train the model and obtain one step prediction.

Since recently edited revisions are more related to the current revision, we give a higher weight to more recent data. We give a final determined predicted value at the t-th week by exponential weighted average (EWA) as:

$$\begin{aligned} P(t) =& \alpha * [P_t + (1 - \alpha)P_{t-1} + (1 - \alpha)^2 P_{t-2} + \ldots + (1 - \alpha)^{t-1} P_1] \\ &\text{s.t.}\quad \alpha \in\ (0, 1] \end{aligned} \tag{2}$$

Here P_t is the ARIMA 1-step prediction value at the t-th week, $P(t)$ denotes the final determined value at the t-th week. The degree of weighting is determined by the hyper parameter α, where setting a larger α will make the model more sensitive to fluctuations. In the t-th week, if the actual revision count deviates from the predicted value beyond the threshold, then we regard that the t-th week is in a burst period. The burst condition is described as follows:

$$\begin{aligned} &\text{if}\quad \text{revision count}(t) \geq \beta * P(t) \\ &\text{then}\quad t \text{ is in burst period} \end{aligned} \tag{3}$$

$B > 1$ is the threshold parameter greater than 1.

3.2 Data Preprocessing

3.2.1 Filtering Low Quality Revisions

In edit history of Wikipedia, we are concerned with revisions that are adding or updating contents, because these contents are highly related with the events that happened at the corresponding time. But there are also a number of revisions that are generated by other reasons. These revisions did not add or update the pivotal content in the page and will have an adverse effect when calculating the TextRank_nfidf value. The following three types of edits of revisions should be removed from our datasets:

1) Minor edits
   Minor edits are usually spelling or grammatical corrections, or formatting that do not change the main content of the pages. These edits will generate a certain number of revisions that store almost the same text page in the database.
2) Robot edits
   Wikipedia conducts regular maintenance to articles and record theses operations as revisions. Robot edits refer to edits done by machine. Robot editors perform adding links, adding tags or reverting revisions from vandalism edits and so on.
3) Vandalism edits
   Vandalism edits include malicious deletions, insulting languages, etc. These revisions will be reverted to the previous revision if they are discovered by other editors.

The above three type of edits of revisions will have a negative impact on our calculation of the net frequency of terms. Figure 2 shows the change of the text length of article "Donald Trump sexual misconduct allegations" over its revision sequence. From the figure we find that since minor revisions and bot revisions only slightly change the text, the term frequency is not changing much during most of minor revisions, eventually reducing the relative net frequencies of important words in the category. Vandalism edits usually delete the whole text, causing the term frequency dramatically dropped in several revisions. But since these changes are abnormal, they should be removed from the edit history.

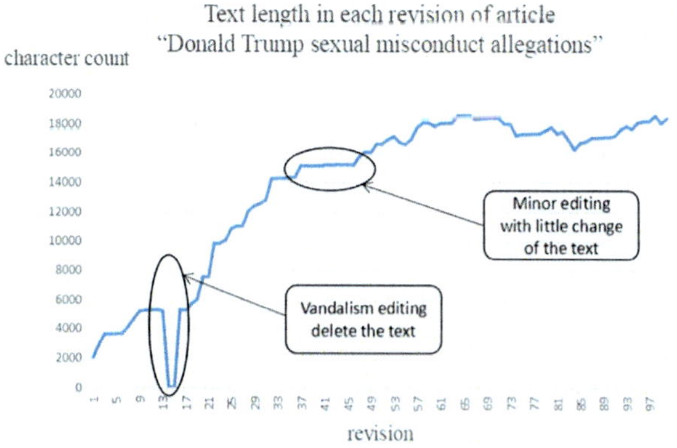

Fig. 2. Character counts in revisions of article "Donald Trump sexual misconduct allegations"

3.2.2 Revision Text Difference (δ *Text*)

The term δ *text* refers to sentences that are added or deleted, or sentences on which words are deleted or added, between two adjacent revisions of one article. There are two main reasons for introducing δ texts.

First, a δ text is likely to be related to event(s) occurred shortly before the time of the δ text. Although spontaneous edits by editors can create new revisions, bursting edits are more likely to be caused as responses to significant event(s).

If only a small portion of an article between two adjacent revisions is modified, most of the article remains unchanged. If we simply use the full text for keyphrase extraction, phrases that were written early in the article will retain high rankings throughout its history, but such phrases are not closely related to significant events. From a δ text, the algorithm can only extract phrases such that an update on word(s) in the same sentence occurred at the time of the δ text.

Secondly, TextRank is a graph-based algorithm that runs in time super-linear to the input text. Since δ texts only use the changed parts between two revisions, reduce the running time can be greatly reduced.

3.2.3 POS Tagger

Noun word groups are meaningful and easy to understand for human. In this paper, we use POS tagger and Chunker provided by Apache openNLP [17] to extract candidate phrases. We apply POS tagging on the revisions of Wikipedia articles to divide each sentence of revisions into chunks, like noun groups, verb groups and so on. The noun POS tagger produces the following classifications:

- NN: Noun, singular or mass
- NNS: Noun, plural
- NNP: Proper noun, singular
- NNPS: Proper noun, plural

The chunks containing noun POS tags are considered as candidate phrases of our keyphrase extraction algorithm, and a graph node is created for each candidate phrase.

3.3 Keyphrase Extraction Based on TextRank_nfidf

TextRank is a well-known method to extract keyphrases. It is inspired from PageRank for web page scoring. TextRank utilizes co-occurrence relationship between words to construct an edge weighted graph. The definition of the score function of TextRank is shown in (4).

$$WS(V_i) = (1-d) + d * \sum_{V_j \in In(V_i)} \frac{w_{ji}}{\sum_{Vk \in Out(V_j)} w_{jk}} WS(V_j) \tag{4}$$

Here, $WS(V_i)$ is the rank score of node V_i, d is the damping factor usually set to 0.85, w_{ji} weight the edge from node V_j to node V_i. $In(V_i)$ is the set of nodes that point to V_i, $Out(V_j)$ is the set of nodes that node V_i points to. w_{ji} is defined as the co-occurrence count of V_j and V_i in a window of maximum W words in the text. TextRank only takes edge weights into account. This leads to a situation such that common words which appear many times in the associated text have a high rank score. In fact, the node weight of V_i denoted by $W(V_i)$ is important in reducing the rank of common words. With node weights, the score function is defined as:

$$WS(V_i) = (1-d) * W(V_i) + d * W(V_i) * \sum_{Vj \in In(V_i)} \frac{w_{ji}}{\sum_{Vk \in Out(V_j)} w_{jk}} WS(V_j) \tag{5}$$

TF-IDF [13], shown below, is a conventional measure for word significance, which can be used as node weights:

$$W(V_i)_{TFIDF} = TF(V_i) * \log_2 \frac{L}{DF(V_i)} \tag{6}$$

Here, L is the number of articles appeared in the history of the given category, and $DF(V_i)$ is the number of articles containing phrase V_i. The number of articles L is changing over time. But for simplicity, we choose L to be the number of all the articles that ever appeared.

In our situation, we want to extract phrases that are frequently edited or newly added during the burst period. Therefore, we focus on the change trend of the phrase frequency around the period. We define the revision term frequency difference (*RTFD*):

$$RTFD_{j,k}(V_i) = TF(V_i, R_{j,k}) - TF(V_i, R_{j-1,k}) \tag{7}$$

Here, $TF(V_i, R_{j,k})$ is the frequency of term V_i in the j-th revision of the k-th article. $RTFD_{j,k}(V_i)$ measures the difference of frequencies of phrase V_i between j-th revision and $(j - 1)$-th revision in the k-th article. We count *RTFD* separately for each article. A positive $RTFD_{j,k}(V_i)$ means the mentions of phrase V_i are increasing. On the contrary, $RTFD_{j,k}(V_i)$ can be negative when deletion happens. In Wikipedia, articles are merged or split, due to duplicate removal or growth of contents. Therefore, monitoring phrase frequency in an individual article does not capture the global trend of the phrase. Thus, we introduce the net frequency (*NF*) to capture net edit activities of each phrase:

$$NF_t(V_i) = \max\left\{0,\ \sum_{k=1}^{L} \sum_{j \in rev(k,t)} RTFD_{j,k}(V_i) \Big/ \sum_{k=1}^{L} |rev(k,t)|\right\} \tag{8}$$

Here, *rev*(*k*, *t*) is the set of revision IDs of the k-th article, created in the t-th week. $NF_t(V_i)$ represents the net frequency of term V_i in the t-th week before the current time. Since adding new contents to an article is more common than deleting contents, the sum of *RTFD*s is usually positive, but the sum can be negative when massive deletes happen. Since negative node weights are not acceptable, we set $NF_t(V_i) \geq 0$, meaning that edit activities of a phrase are reflected only when its net phrase frequency is increasing.

We find that even during a burst period, the net frequency of a phrase in each week still jitters around the period. Therefore we apply the exponential moving average to smooth the net frequency, as follows:

$$NF(V_i) = \frac{NF_1(V_i) + \eta * NF_2(V_i) + \eta^2 * NF_3(V_i) + \eta^3 * NF_4(V_i)\ldots}{1 + \eta + \eta^2 + \eta^3 + \ldots} \tag{9}$$

Here, $NF_1(V_i)$ is the net frequency of node V_i at the current week, $NF_2(V_i)$ is the net frequency of node V_i at the last week and so on. The coefficient $0 \leq \eta < 1$ gives a damping factor. Now the *nfidf* node weight is given as:

$$W(V_i)_{nfidf} = NF(V_i) * \log_2 \frac{L}{DF(V_i)} \tag{10}$$

4 Experiments and Evaluations

4.1 Datasets

To evaluate our method, we use all the articles in the Wikipedia category "Political scandals in the United States" and its subcategories. We collected their revision histories from 2014/3/12 to 2018/8/2. 288 articles were edited at least once during the period and 32618 revisions were created. The number of revisions of these articles ranges from 1 to 1700. 82.9% of the articles had less than 100 revisions, 14.3% articles had 100–500 revisions, and 2.8% of them had over 500 revisions.

4.2 Results on Burst Period Detection

After removing minor, bot and vandalism revisions, we apply our ARIMA_EWA method to the time series, where we set $\beta = 2$, and set $\alpha = 0.33$ and 1. Figure 3 shows the burst detection result under different α. The blue bar is the weekly revision count. The grey area is the predicted curve generated by ARIMA_EWA. The orange bars are the burst weeks in which the revision count deviates from the predicted value. From Fig. 3 we can see that:

1) ARIMA_EWA performs well in detecting outliers from the time series of revision counts.

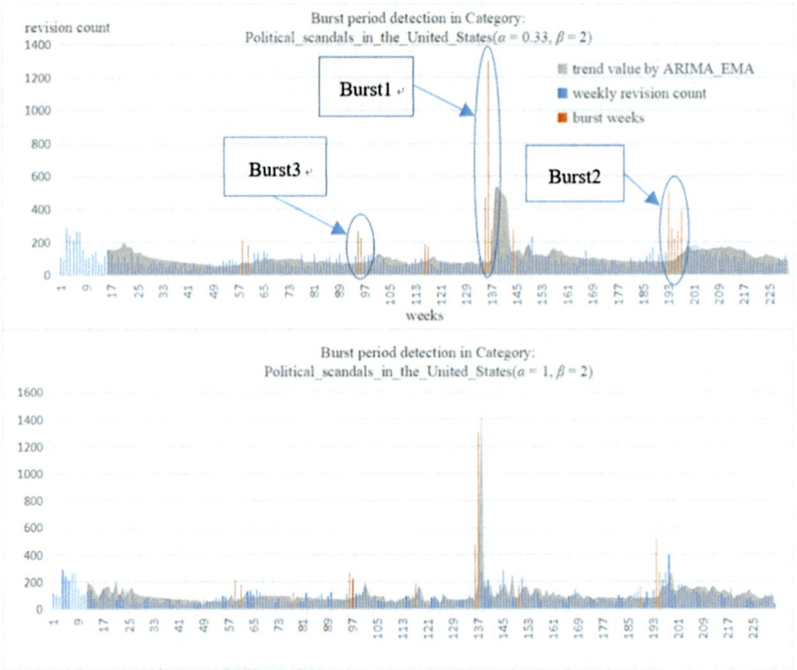

Fig. 3. Burst period of edits on category "Political scandals in the United States" with varying α

Table 1. Top 3 burst periods

	Burst1	Burst2	Burst3
From	2016/10/26	2017/12/6	2016/1/20
To	2016/11/16	2018/1/10	2016/2/3
Sequence	135th–137th weeks	193th–197th weeks	95th–96th weeks

2) Setting a larger α will make the model more sensitive to fluctuations, leading to the shorter burst period. On the other hand, setting a smaller α makes the prediction curve smoother, thus extending the burst period.

Table 1 shows the top-3 most intense burst periods (based on area, $\alpha = 0.33$, $\beta = 2$).

4.3 Results of Keyphrase Extraction

4.3.1 Comparison by Weekly Average Edit Frequency

We apply TextRank and TextRank_nfidf to Burst1, where we set $\eta = 0.8$. The top 10 phrases ranked by each method are shown in Table 2.

Table 2. Top ranked 10 phrases in Burst1

TextRank				TextRank_nfidf			
1	Trump	6	election	1	Trump	6	**harassment**
2	Clinton	7	**United States**	2	Clinton	7	misconduct
3	campaign	8	misconduct	3	sexual assault	8	**debate**
4	sexual assault	9	**president**	4	election	9	**lawsuit**
5	**case**	10	**time**	5	campaign	10	**abuse**

Comparing the results of the two methods, we notice that TextRank is promoting common words like "case" and "time," because these words are co-occurring with bursty keywords and then receive a higher rank. But in TextRank_nfidf, which reflects edit frequency as node weights, these words were removed due to low nf-idf values. Figures 4 and 5 show the difference in weekly average edit frequencies between common phrases and specific phrases selected by TextRank and TextRank_nfidf. Phrases like "president" and "United States" appear in many articles in the category. In fact, these phrases are frequently used and have weak relationship with outbreak of a real event. TextRank_nfidf successfully removes these phrases. Certain specific phrases like "Trump" and "sexual assault" were triggered by real events. Then they have higher average edit frequencies only in several weeks. These phrases have strong relationship with the real events in the corresponding weeks.

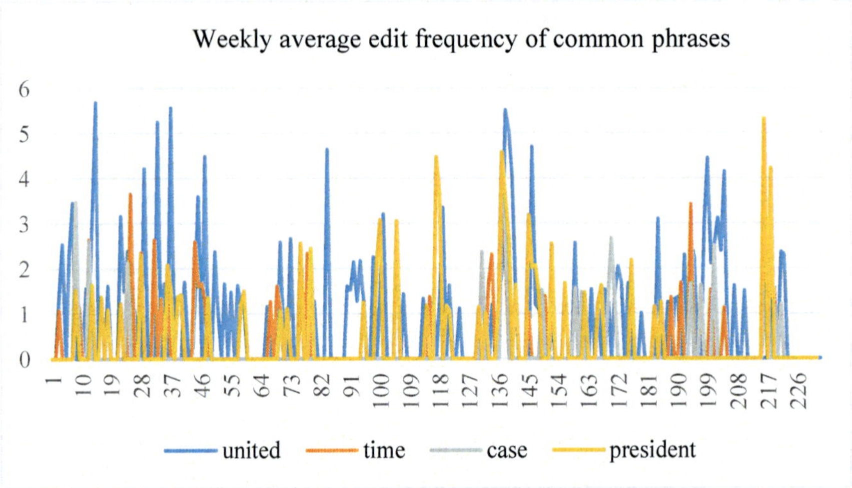

Fig. 4. Weekly average edit frequencies of common phrases

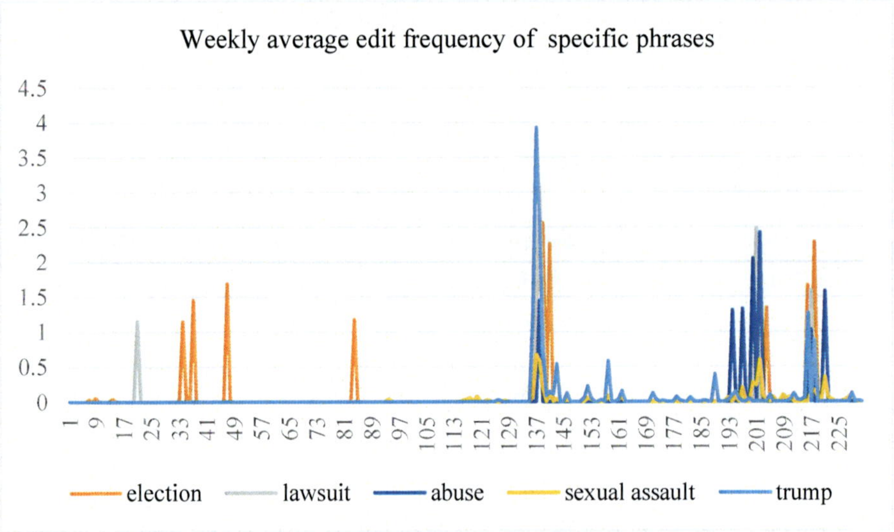

Fig. 5. Weekly average edit frequency of specific phrases

4.3.2 Comparison by Weekly TextRank_nfidf Score

Our TextRank_nfidf algorithm is independent of the burst period detection algorithm. In addition to extracting keyphrases during the burst periods, we can also apply this method to obtain weekly TextRank_nfidf score for each phrase, for observing the long-term editing trend of highly-ranked phrases throughout the whole edit history.

Figure 6 shows the effect of changing parameter η on the weekly scores of specific phrase "lawsuit." When $\eta = 0$, which means no smoothing effect, the curve shows high

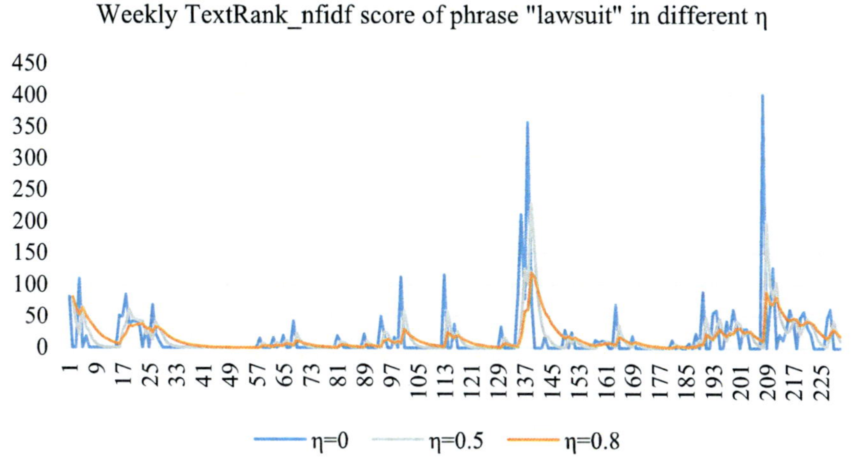

Fig. 6. Weekly score curve of specific phrase in different η

TextRank_nfidf score in certain weeks, but the jittering of the scores is still intense during these weeks. Next, when η is 0.5 and 0.8, then the curve becomes smoother and the long-term editing trend of phrase is also easy to observe. In the subsequent experiments, we set the parameter η to be 0.8 to show the long-term editing trend of the phrases.

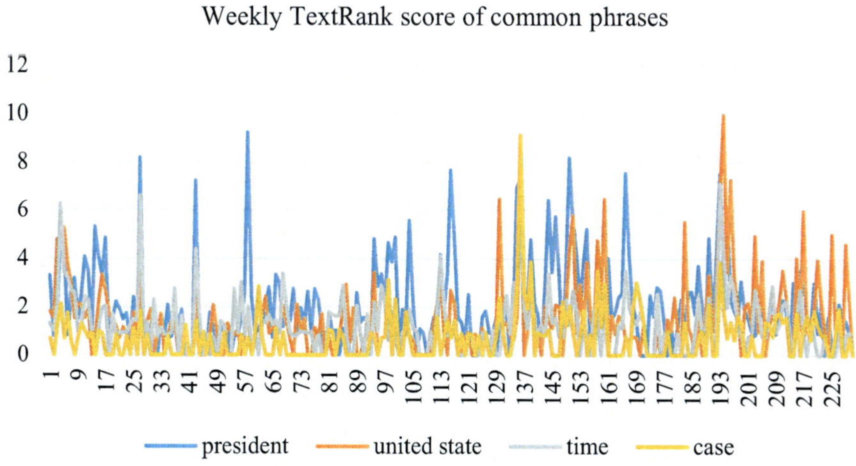

Fig. 7. Weekly TextRank score of common phrases

Next, we show the weekly scores by the two methods to compare long term trends of phrases. Figures 7 and 8 show the difference in weekly scores between common phrases and specific phrases selected by TextRank and TextRank_nfidf. Compared with

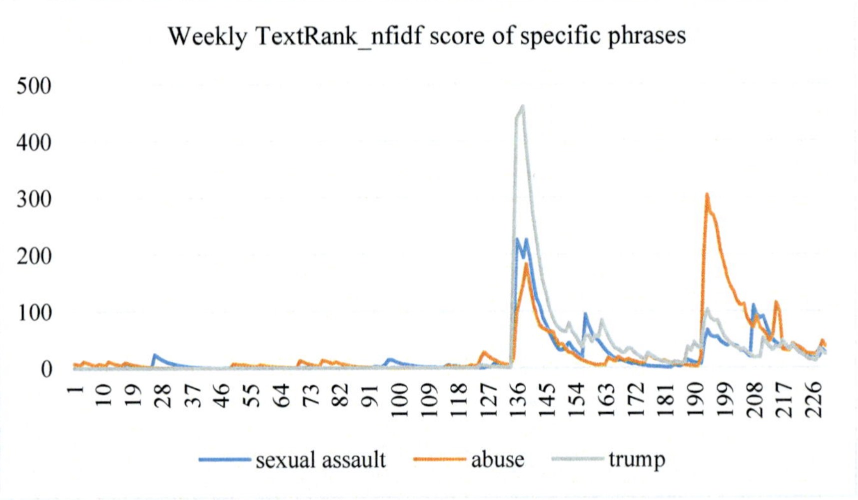

Fig. 8. Weekly TextRank_nfidf score of specific phrases

the results of weekly average edit frequencies, we find that weekly scores provide a more intuitive and convincing presentation of phrase editing trends. TextRank_nfidf successfully removes the phrases like "president," "United States" due to their lower nfidf values. These phrases appear in many articles in the category and have weak relationship with development of a real event. But in TextRank, these phrases still have high ranks. Certain specific phrases like "Trump" and "sexual assault" were triggered by real events. TextRank_nfidf successfully gives high ranks to these phrases in their bursting weeks.

4.3.3 Compared with Google Trends

Now we apply our method to the other two different burst periods. Table 3 lists the top 10 phrases in Burst1 and Burst2. The keyphrases extracted from these two periods are apparently different, indicating that these phrases are related to distinct real-life events. We select two phrases from each burst and compare them with Google Trends to examine whether these phrases correspond with real-life events around the burst periods. Figures 9 and 10 show the comparison between weekly TexRank_nfdif score of the phrases and search trends given by Google Trends.

Overall, the extracted keyphrases have similar temporal patterns with the results by Google Trends, indicating validity of our work. The first burst represents the Trump sexual misconduct allegations during the 2016 US president election. The second burst is related to the Weinstein effect. The latter involves allegations against politicians including Roy Moore.

In addition to similar temporal patterns, our method can give a better tracking of editor's attentions in edit history. Figure 10 shows phrase "Clinton" as an example. Compared with Google Trends, our method detects a spike around the 193th week, because the event "Weinstein effect" appeared during this period. Although this event is

Table 3. Top ranked 10 phrases in Burst1 & Burst2

Burst1 (2016/10/26–2016/11/16)				Burst2 (2017/12/6–2018/1/10)			
1	**Trump**	6	harassment	1	**Roy Moore**	6	harassment
2	**Clinton**	7	misconduct	2	**Hollywood**	7	misconduct
3	**sexual assault**	8	**debate**	3	**Weinstein**	8	lawsuit
4	**election**	9	lawsuit	4	abuse	9	**December**
5	**campaign**	10	abuse	5	**senator**	10	**effect**

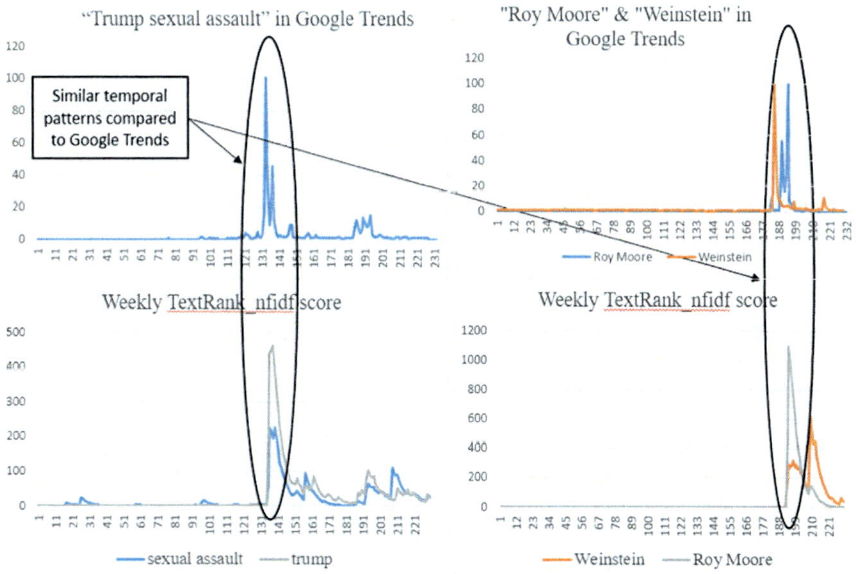

Fig. 9. Similar temporal patterns compared to Google Trends results

not largely related with Hillary Clinton and Bill Clinton, editors checked sexual scandal-related articles in the category, and then re-edited the articles about Hillary Clinton's misconduct allegations in 2016 US president election. Thus the edit activity shows a peak around the 193th week, but no corresponding search heat appears in Google Trends.

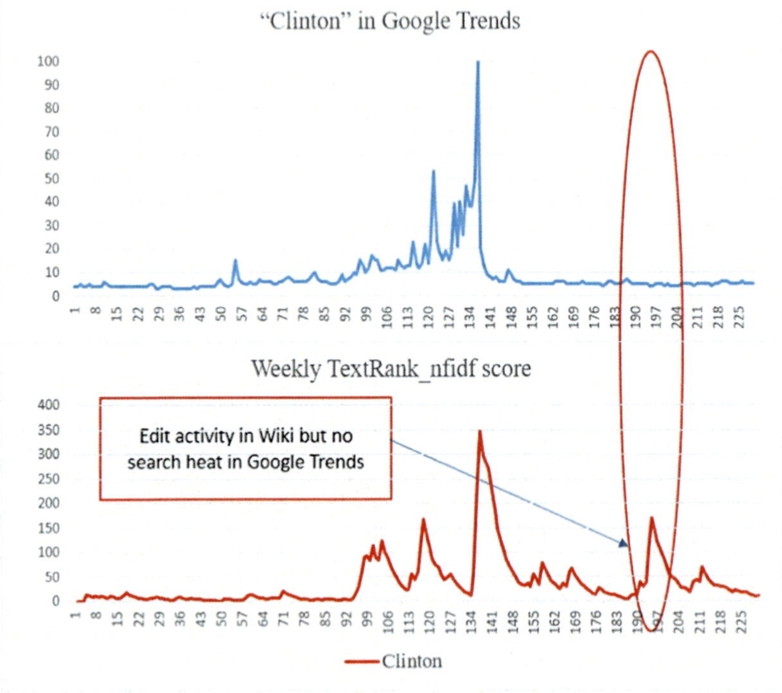

Fig. 10. Difference of temporal patterns compared to Google Trends results

5 Migration of Editing Activities Between Article Categories

In this section, we further examine characteristics of TextRank_ndfidf score, with focus on how the given article set affects keyphrase extraction, and through case studies we show that edit activities on a certain topic are migrating over article categories.

In the extraction of keyphrase from the dataset "Political scandals in the United States," we found that several burst periods were related to real events on sexual harassment. In order to find more articles frequently edited on these events, we collected all the articles in category "Sexual harassment" and its subcategories, and apply the same time span as "Political scandals in the United States." 125 articles were edited at least once during the period, and 29002 revisions were produced. Table 4 shows the statistics of the article set and Fig. 11 shows the weekly revision counts of the two article sets, and the real-life events corresponding to the burst periods.

Table 4. Information about two article sets

Article sets	Time span	# of articles	# of revisions
Political scandals in the United States	2014/3/12–2018/8/2	288	32618
Sexual harassment	2014/3/12–2018/8/2	125	29002

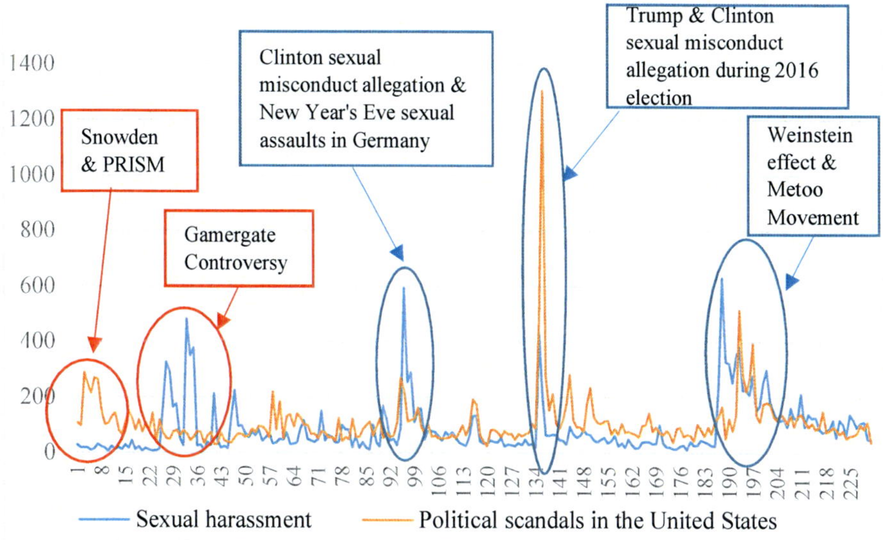

Fig. 11. Weekly revision counts of two article sets

The blue circles represent burst patterns similar between the two article sets, while the two red circles indicate major burst patterns disagreeing with each other. One is the Snowden event in January 2014 in category "Political scandals in the United States," which is a scandal unrelated to sexual harassment. The other is Gamergate Controversy in August 2014, which is a sexual harassment case in video game industry.

Now we apply our burst detection method to the Sexual harassment article set and apply TextRank_nfidf to obtain keyphrases in each burst period. Table 5 shows the top-ranked phrases in each burst period. The keyphrases extracted in each period are well representing the corresponding events that occurred in the period.

The first burst period is about "Gamergate Controversy," the harassment campaign targeted at several women in the video game industry, notably game developers Zoë and Quinn. Then the controversy extends to the harassment campaign conducted primarily through the use of the hashtag #GamerGate.

The second burst period contains two events. One is "Clinton sexual misconduct allegation," and the other is "New Year's Eve sexual assaults in Germany." These two events do not have much relationship; they just coincidentally happened around the same time period. But the related articles of these two events are both frequently edited in this period and detected by our algorithm. We can isolate two events by grouping checking phrases by co-occurrence in revisions and link structures.

The third burst period is about Trump's and Bill Clinton's sexual misconduct allegations during the 2016 president election, involving two major events: "Donald Trump Access Hollywood tape" and "Bill Clinton's sexual scandal with Lewinsky in 1990s."

In Fig. 12, we can see that top phrases during this burst period can be clearly divided into two groups based on their long-term editing trends. Group 1 contains words "Lewinsky," "Broaddrick," "Clinton," all of which are persons related to the Clinton sexual

Table 5. Top phrases in each burst period of article set "Sexual harassment"

Event in category "Sexual harassment"	Burst periods	Keyphrases extracted by TextRank_nfidf
Gamergate Controversy in 2014.8	25th week–30th week/2014.9.3–2014.10.8	game, Zoë, Quinn, gamergate, controversy, video, transport, august ...
	32th week–34th week/2014.10.22–2014.11.5	gamergate, game, controversies, hashtag, Quinn, Assange, movement ...
Clinton sexual misconduct allegation & New Year's Eve sexual assaults in Germany	95th week–100th week/2016.1.6–2016.2.10	Cologne, Clinton, Germany, police, taharrush, jamai, square, Egypt, düsseldorf...
Trump & Clinton sexual misconduct allegation during 2016 president election	134th week–135th week/2016.10.5–2016.10.12	Clinton, Trump, Lewinsky, Broaddrick, October, audio, tape, Donald...
Weinstein effect & Metoo movement	186th week–192th week/2017.10.4–2017.11.15	Weinstein, Harvey, october, actress, Clinton, Hollywood, casting, film...
	201th week – 203th week/2018.1.17 –2018.1.31	Metoo, Weinstein, landis, misconduct, Italy, actress, hashtag, Vairamuthu, movement, allegations...

misconduct. Group 2 contains "Trump," "Audio," and "Tape," which are related to event "Trump Access Hollywood tape." In addition, the following phenomena are observed:

1) Articles about Bill Clinton's sexual misconduct have been edited with bursts throughout their edit histories. This is because the scandal is a long-lasting process, where related news is constantly appearing, causing occasional edit bursts.
2) Trump's editing burst first appeared in the 134th week (Oct. 2016). The news around the burst period indicates that two days before Trump's second presidential debate in 2016, the Washington Post published an audio clip, which causes creation of an article titled "Trump Access Hollywood tape".

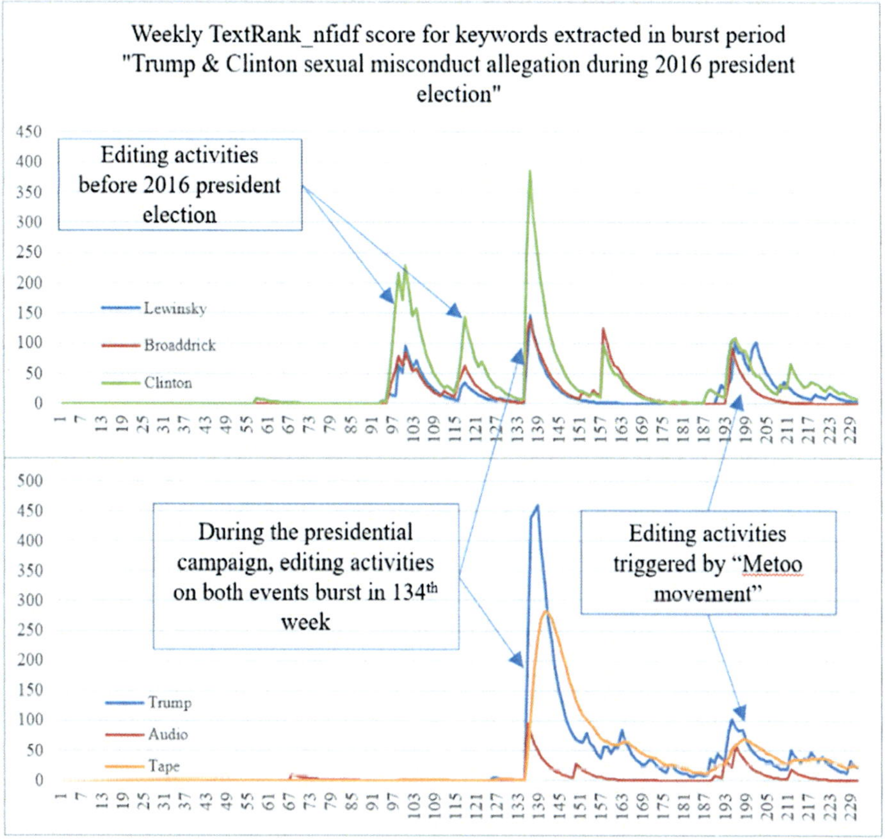

Fig. 12. Weekly TextRank_nfidf score for keywords extracted in burst periods of "Trump & Clinton sexual misconduct allegation during 2016 president election"

3) In the 193th week, both phrase groups have a low-level editing burst. This is because "Metoo movement" occurred at this time, which has a broad social impact, creating articles and their revisions regarding the Metoo movement. Existing articles "Bill Clinton sexual misconduct allegations" were also added with reference to the Metoo movement.

In summary, bursts on articles are often triggered by real-life events, including news release. However, edit bursts are widely different in scale, from events localized to a small article set, such as (1) and (2) above, to wide-scale events, such as 3).

Next we present an interesting phenomenon where edit activities are migrating between categories, as the event evolves. Weekly TextRank_nfidf curves of keyphrases are appropriate for tracing migration of editing activities on different categories in Wikipedia.

Here, we take the event "Weinstein effect & Metoo movement" as an example. Figure 13 shows weekly TextRank_nfidf scores of four phrases. First in October 2017, sexual abuse allegations against Weinstein happened and article "Harvey Weinstein" started being frequently edited, because this is the only article in the category of "Sexual harassment." Around the 186th week, only the category of sexual harassment has high-level edit activities, and no edit activities in "Political scandals in the United States." Then as the event evolves, phrase "Weinstein effect" started to be used to refer to a worldwide wave of allegations on powerful men of sexual misconduct that began in the United States in October 2017. Then new article "Weinstein effect" was created in category "Political scandals in the United States" in November 2017, and phrase "Weinstein" began highly edited at the 191st week in this category.

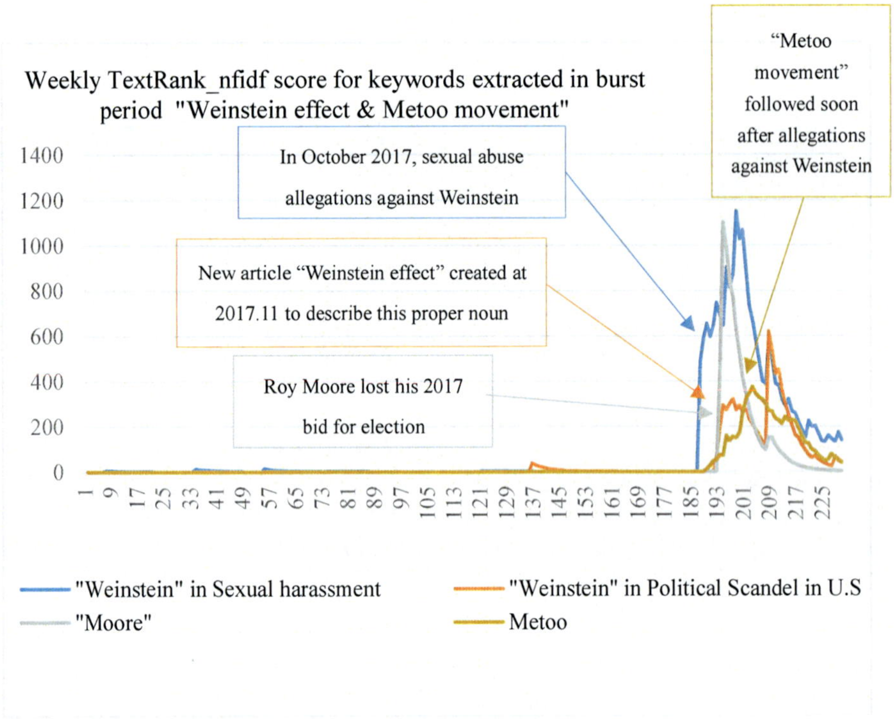

Fig. 13. Migration of editing activities between categories

Influenced by "Weinstein effect," Roy Moore was accused by three women and lost his 2017 bid for election to the United States Senate. Article "Roy Moore" began to be frequently edited, reflected in the weekly score curve around the 195th week. About one month after the sexual abuse allegations against Harvey Weinstein, the Metoo movement broke out. The editing activities on phrase "Metoo" show a peak around the 204th week, having approximately from weeks delay from the peak of editing activities of phrase "Weinstein" in category "Sexual harassment."

Figure 14 shows the weekly changes of the trends of phrases "Weinstein" and "Metoo" by Google Trends. In comparison with Fig. 13, we can point out our weekly score curves have the following differences from Google Trends.

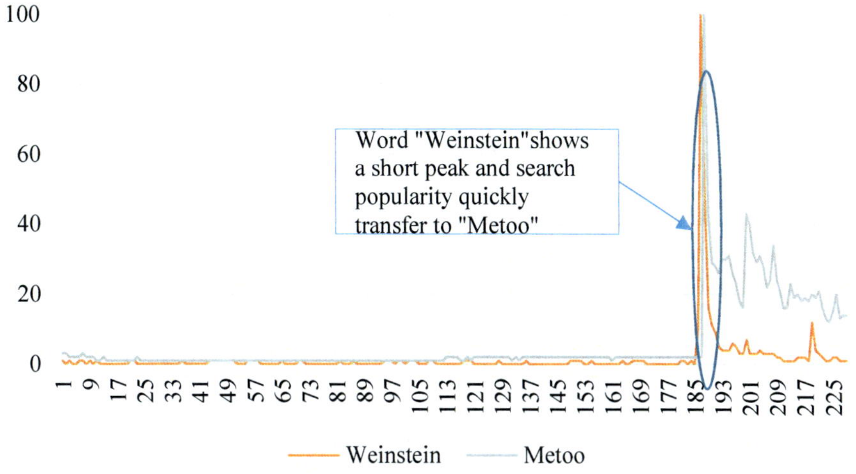

Fig. 14. "Weinstein" & "Metoo" in Google Trends on burst period of "Weinstein effect & Metoo movement"

1) Google trends shows a short peak of "Weinstein" which quickly returns to normal around the 183th week. Then the search heat is shifted to "Metoo." But in our method, word "Weinstein" shows continuous editing activities in two article sets which lasted for more than 20 weeks. This is because Google Trends is focusing on Internet user's search popularity, which can be transient. Since user's interest quickly shifted from "Weinstein effect" to "Metoo movement," word "Weinstein" only had a short-lived search peak. On the other hand, Wikipedia is essentially a knowledge database, where editors need to add new information related to "Weinstein," so although the search heat has shifted to "Metoo movement," the editing activity about "Weinstein" continued.
2) The result of Google Trends gives an overall search popularity of phrases by Internet users. On the contrary, our approach detects edit activities nearby target phrases in a given article set. By comparing the score curves of the same phrase over different article sets, we can discern on which article sets edits are active. Also, migration of edit activities between related phrases can be detected. The underling article sets, which can be collected from a category, indicate the contexts of the phrases. On the contrary, in Google Trends, adding necessary words for clarifying contexts often result in failure of detecting any burst, due to insufficiency of documents.

The above case studies show that TextRank_nfidf can find significant phrases receiving intense edit activities, from the entire edit history. Our method can find actively edited phrases from relatively small sets of articles, which are advantageous in clarifying the contexts of bursts, localizing subsets where bursts are occurring, as well as tracking how bursting edits are migrating. Specifying appropriate article sets are important for our approach. We believe that we can find articles related to a burst or phrase of interest, through word co-occurrences, exploration of link and category structures, combined with burst patterns.

6 Conclusion and Future Work

In this paper, we proposed a novel approach for extracting keyphrases from Wikipedia articles that are significantly related to bursting edits. We proposed TextRank_nfidf which utilizes both edge weights and node weights, and reflects smoothed difference of phrase frequencies between revisions as node weights. Our new method can effectively extract bursty keyphrases around which new descriptions are intensively added, rather than extracting keyphrases having high public attention. Our experiments show that weekly TextRank_nfidf score curves indicate clear response to significant real-life events, and succeeding editing activities. Also, our approach can show on which article sets bursting edits are happening. Research so far can be applied in domains like edit assistance, event causality analysis and attention trend analysis.

There are still rooms for improvement to our proposed method. Currently, our method extracts phrases around which bursting edits are happening, from a given article set. In future work, we explore visualization of the node graph of TextRank_nfidf, for observing the temporal change of neighbors of keyphrases. Also, updating article sets for detecting migration of editing activities should be explored.

References

1. Al Tamime, R., Giordano, R., Hall, W.: Observing burstiness in Wikipedia articles during new disease outbreaks. In: Proceedings of the 10th Web Science, pp. 117–126. ACM (2018)
2. Bellaachia, A., Al-Dhelaan, M.: NE-Rank: a novel graph-based keyphrase extraction in Twitter. In: Proceedings of the Conference on Web Intelligence & Intelligent Agent Technology, pp. 372–379 (2012)
3. Box, G., Jenkins, G.: Time Series Analysis: Forecasting and Control. Holden-Day, San Francisco (1970)
4. Chen, Z., Iwaihara, M.: Detection of bursty and significant keyphrases from Wikipedia edit history. In: 6th IEEE International Conference on Big Data and Smart Computing (BigComp 2019), Kyoto, pp. 99–103, February 2019
5. Christiansen, L., Schimoler, T., Burke, R., Mobasher, B.: Modeling topic trends on the social web using temporal signatures. In: Proceedings of the 12th International Workshop on Web Information and Data Management, pp. 3–10, November 2012
6. Han, J., Pei, J., Yin, Y.: Mining frequent patterns without candidate generation. ACM SIGMOD Rec. **29**(2), 1–12 (2000)
7. He, D., Parker, D.S.: Topic dynamics: an alternative model of bursts in streams of topics. In: Proceedings of the 16th ACM SIGKDD on Knowledge Discovery and Data Mining, pp. 443–452, July 2010
8. Lau, J.H., Newman, D., Karimi, S., Baldwin, T.: Best topic word selection for topic labelling. In: Proceedings of the 23rd International Conference on Computational Linguistics, Posters, pp. 605–613 (2010)
9. Liu, R., et al.: Mining phase evolution for hot topics: a case study from multiple social media platforms. In: IEEE International Conference on Systems, Man, and Cybernetics (SMC), pp. 2814–2819 (2017)

10. Mihalcea, R., Tarau, P.: TextRank: bringing order into text. In: Proceedings of the Conference on Empirical Methods in Natural Language Processing (2004)
11. Page, L., et al.: The PageRank citation ranking: bringing order to the web. Stanford InfoLab (1999)
12. Yu, Q., Jibin, L., Jiang, L.: An improved ARIMA-based traffic anomaly detection algorithm for wireless sensor networks. Int. J. Distrib. Sens. Netw. **2016**, 1–9 (2016)
13. Ramos, J.: Using TF-IDF to determine word relevance in document queries. In: Proceedings of the 1st Instructional Conference on Machine Learning, vol. 242, pp. 133–142 (2003)
14. Tsuruoka, Y., et al.: Developing a robust part-of-speech tagger for biomedical text. In: Panhellenic Conference on Informatics, pp. 382–392 (2005)
15. Zesch, T., Gurevych, I., Mühlhäuser, M.: Analyzing and accessing Wikipedia as a lexical semantic resource. In: Data Structures for Linguistic Resources and Applications, pp. 197–205 (2007)
16. Liu, Y., Gao, Z., Iwaihara, M.: Identifying evolutionary topic temporal patterns based on bursty phrase clustering. In: Asia-Pacific Web (APWeb) and Web-Age Information Management (WAIM) Joint Conference on Web and Big Data, pp. 276–284. Springer, Cham (2017)
17. http://opennlp.apache.org/

Emotion Detection on Twitter Textual Data

Fan Jiang(✉) and Colton Aarts

Department of Computer Science, University of Northern British Columbia,
Prince George, BC, Canada
{Fan.Jiang,aarts}@unbc.ca

Abstract. Textual data is one of the most important types of data nowadays. With the use of social media like Twitter and Facebook, the amount of textual data is growing at an unimaginable speed. To fully use these data, it is very important to extract useful information from it as much as possible. While some information can be directly extracted from the text, other information needs to be discovered with an advanced method. For instance, the feelings of authors while the text was written, or the emotions that authors wanted to express. This leads to the use of sentiment analysis. Recently, sentiment analysis has drawn attention from many researchers. With the growth of online resources of textual data, the ability to detect different emotions from textual data is becoming very useful. In this paper, we present an application that can detect four distinct emotions from Twitter posts. They are joy, anger, fear, and sadness. The evaluation result shows the accuracy and applicability of our application.

Keywords: Sentiment analysis · Big data · Social media analysis · Emotion detection

1 Introduction

The roots of sentiment analysis can be tied back to work done in [1,12,14]. In these papers, the authors started exploring the idea of sentiment analysis. In [12], the authors look into determining if reviews of a movie are positive or negative. They compare a number of different techniques, namely support vector machines, Naive Bayes classifiers, and maximum entropy classifiers. In [14], the author approaches the same problem as the authors of [12] but uses a different technique. The technique used in [14] uses part of speech tagging and individual word weights to produce positive or negative labels. In [1], the authors move from simple positive/negative reviews and ratings to a range based rating system. They test a number of different methods and discover a number of new problems that arise from a problem with non-binary labels.

As the popularity of sentiment analysis continued to grow, it became of interest to researchers to be able to work in new domains. Furthermore, researchers

W. Lee et al. (Eds.): *Big Data Analyses, Services, and Smart Data*, AISC 899, pp. 66–76, 2021.
https://doi.org/10.1007/978-981-15-8731-3_5

are no longer satisfied by simply detecting just positive and negative sentiment. In order to expand both the number of domains and the range of emotions capable of being detected, the creation of annotated training and testing collections was required. The creation of one of these corpora is outlined in [18]. This paper details the creation of a large collection of annotated text that contains a vast amount of information manually extracted from large text pieces. The authors outline the process in which they obtained the labels and explained what the different fields mean in their corpus. This corpus allowed the application of sentiment analysis to a more diverse domain as it was no longer dependent on online reviews. The authors in [11] describe a method of collecting a corpus from Twitter that contains positive, negative, and neutral tweets. The authors determine if a tweet is positive or negative based on the presence of different emoticons. The objective corpus was collected from a number of news reporting accounts. With the introduction of a way to automatically collect data from a social media website, the authors of [11] provided future researchers with an easy way to access the data that is contained on these sites. In [9], the creation of a corpus of annotated Twitter posts with four different emotions labeled: anger, fear, joy, and sadness. It contains the labeled text as well as the level of the emotion that is being expressed as well. This corpus has allowed other authors to start classifying text into emotions that are more distinct than positive and negative.

There are a variety of different techniques that are used for the classification of information. In [4], the authors propose a lexical system to classify tweets. In [19], the authors give a review of the different deep learning and neural networks approaches that have been used in sentiment analysis. The authors in [5] use both k-nearest neighbors and support vector machines to create a pair of approaches for the labeling of tweets. Ensemble classifiers are used in [10]. They used a deep learning technique in conjunction with surface-level techniques to create an ensemble to classify tweets.

In this paper, we propose an application that can perform sentiment analysis on Twitter text data. Our application creates an ensemble classifier built from a support vector machine, a lexical classifier, and a Naive Bayes classifier.

The remainder of this paper is organized as follows. The next section will cover the background studies, in which we will introduce some important existing theories and methods that can be applied to sentiment analysis. Section 3 will present our social media text sentiment analysis application by walking through examples. In Section 4, we present the outcome of the application and evaluation of this outcome. In the end, a conclusion will be presented.

2 Backgrounds and Related Works

The process of sentiment analysis can be broken down into two basic steps: preprocessing and classification. Each of these steps can be further broken down into detailed steps. In general, preprocessing is where the documents are taken in and transformed from whatever written language. They were collected into a form that a computer can more easily understand and work with. In most cases, this new form is some numerical representation, usually appears as a matrix. After this transformation has taken place, the matrix will then be given over to the classification step. That is where the document will actually be given a label.

2.1 Preprocessing

In our application, several techniques are used for preprocessing, and they are: stop word removal, stemming, conversion to a Term-Frequency Inverse Document Frequency (TF-IDF) matrix, and then dimension reduction using Singular Value Decomposition (SVD).

Stop Word Removal. Stop words are words that have been identified as words that are common in regular speech and that their presence in a sentence does not influence the content or sentiment of the sentence. The removal of these words helps reduce the overall complexity of the classifier as well as improving the recall of the classifier [17]. The list of stop words that we are using was obtained from [13]. This is a list of English words that occur in a large number of sentences, and they have been determined to have no effect on the overall contents of the sentence. The process of removing stop words is fairly simple. The stop words are held in a set. Each document is split into single words. These words are then checked against the set of stop words. If a word is found to be in the set, it is removed from the list of words that belongs to the document. The removal of stop words helps control the overall dimensions of the search space for the classifier. This, in turn, will make the classifier more efficient and more accurate.

Stemming. Stemming a word is the process that is used to return the word to its root or stem word. An example of this would be to take the word “running” and to change it to the word “run”. The steaming process that is used in this application is from the NLTK package [7]. This package calls a method from the Wordnet package [6]. This method works by utilizing two methods. The first of which is that Wordnet has a list of endings that can be attached to words. Each word passed into the method is analyzed to determine if it contains one of the endings. If it does, then ending is removed and reconstruction rules are utilized to return a word that is close to the intended root word. The second method is a list of possible exceptions to the first list. Wordnet strives to utilize these two rules in an intelligent fashion to stem the supplied words in a logical fashion [6].

Term-Frequency Inverse-Document Frequency. Converting the text into a matrix representation is a standard way to represent the material to the computer. One of the common matrix representations used is the TF-IDF matrix. This matrix represents the occurrence of each word in a document against how often the word occurs across all documents. Words that occur in almost all documents are given lower values as they are deemed to be less important than words that occur in a select few documents. We are using scikit-learn's TF-IDF Vectorizor. This vectorizor will create a matrix where each row is a term, and each column represents the different documents [13]. The equation used to calculate each term-document pair is as follows:

$$tfidf(t, d) = tf(t, d) \times idf(t) \tag{1}$$

Where t is the term, d is the document, $tf(t, d)$ is how many time the term t occurs in document d and

$$idf(t) = log(\frac{1+n}{1+df(t)}) + 1 \tag{2}$$

where n is the total number of documents and $df(t)$ is the number of documents that contain the term t. The ones that are added to the numerator and denominator represent an additional imaginary document that contains all terms exactly once. This is done to prevent division by zero. The one that is added at the end of the equation is used to make sure that terms that occur in every document and not entirely ignored. This is important in our context because we have already removed the stop words at this point, and we want to make sure that all the remaining words are at least available to the classifiers for consideration. The creation of a TF-IDF matrix helps the classifier understand which words are potentially more important than others. However, one of the problems that can arise from using this matrix is that there can be thousands of different words in a corpus. This makes the dimensions of the TF-IDF matrix too large to use by itself. To handle this problem, it is common to implement dimension reduction techniques like SVD.

Singular Value Decomposition. SVD is the process of reducing the number of dimensions that are present in a matrix. This process has been shown to combine words that have been used in similar contexts into a single value [16]. Reducing the number of dimensions (or the rank of the matrix) is useful for sentiment analysis because, in even a small number of sample documents, the rank of the matrix can be incredibly large. This can lead to long processing times and the possibility of losing accuracy to unneeded information.

The process of reducing the rank of a matrix using SVD utilizes the fact that a matrix C of dimensions $M \times N$ can be represented as

$$C = U\Sigma V^T \tag{3}$$

where U is the $M \times M$ matrix whose columns are the orthogonal eigenvectors of CC^T, and V is the $N \times N$ matrix whose columns are the orthogonal eigenvectors of C^TC [16]. The matrix Σ is the $M \times N$ matrix comprised of the singular values. Singular values are the ordered square roots of the eigenvalues from the matrix CC^T. In order to reduce the rank of the final output matrix, a person selects a number of the singular values k to keep. You then dispose of all values of Σ that are lower than k and you construct the final matrix $C_k = U\Sigma_k V^T$ [16]. The value of k is important. It is common to choose a value in the hundreds. For our application, we have chosen 800 for this value. This gives us a reasonable number of dimensions while still retaining a good amount of the original information.

2.2 Classification

After the text has been through the preprocessing stage, it is ready to be given to the classifier. The classifier will take in the text and will return a single value determining what sentiment the text is expressing. In our application, we have created an ensemble classifier from an SVM, a lexical classifier, and a Naive Bayes classifier.

Ensemble Classifiers. Ensemble classifiers are a group of classifiers that are used together to classify a piece of text. The reason to use an ensemble is that you are hoping that the different classifiers will cover the individual weaknesses that they could exhibit alone [3]. In our application, we are using three different classifiers and taking the majority vote from the results. In the case of a nonbinary set of labels, if there is no agreement amongst the classifiers, we will give the document a label of 0.

Support Vector Machines. SVMs are a classifier that is generally used to classify two distinct groups. They do this by constructing a hyperplane that separates the two groups with a maximized margin [2]. This margin is the distance from a point belonging to one of the classes to the plane. For a hyperplane to be considered valuable, it has to maximize the size of this margin for points in both classes [2]. In order to classify multiple different classes, SVMs use one of two different approaches. The first is a one-against-one approach. This approach constructs a classifier for each pair of classes. This means that in the case of n classes it will construct $\frac{n \times (n-1)}{2}$. The other option for multiclass classification is a one-against-all approach. In this approach, a classifier is constructed that compares a class against all other classes. This means that there will only be n classes created. Both of these strategies are valid, and choosing the correct one greatly depends on the problem at hand [8]. The implementation that we are using in this application is from the scikit-learn Python package [13]. For its support vector classifier (SVC), the one-against-one approach is used [13].

Lexical Classifier. Lexical classifiers are one of the simpler classifier types. They work from a supplied list of emotional words. These words need to be labeled with the corresponding emotions as well as an associated intensity. Using this list of words, the classifier will determine the emotion of a document according to the presence and number of the words that occur in the list. There are different ways to determine the final label for a document. One approach is to sum all the different emotional weights and then return the emotion that has the largest score.

Naive Bayes. Naive Bayes classifiers work under the assumption that there is conditional independence between every pair of features [13]. For document classification, this means that all the words are independent of the words that occur around them. In a bag-of-words model, this assumption has already been made, so Naive Bayes is a good choice. There are many different types of Naive Bayes classifiers to choose from. One of the most popular is the Complement Naive Bayes [13]. It has been shown to outperform the other popular Naive Bayes classifiers while still be fast and relatively easy to implement [13]. This form of Naive Bayes classifier works by implementing the standard multi-nominal Naive Bayes as well as including information regarding the complement of each class [13,15]. It has been shown that the Complement Naive Bayes classifier works better than the Multi-nominal classifier, especially when dealing with imbalanced data sets [15].

3 Our Approach Walk Through

Our application takes in two parameters. One is a training file, and the other is the file you want to classify based on the training file. In this section, we will use the following sentence label pairs for training in Table 1:

Table 1. Training sentences and their emotion labels

SID	Training sentences	Emotion labels
1	I am so happy and excited, I won the lottery!	JOY
2	Today was a good day	JOY
3	I went shopping today and the grocery store was so busy. There were people everywhere and I was getting annoyed	ANGER
4	Ahhh I am terrified of thunderstorms	FEAR
5	All this rain is really depressing	SADNESS
6	If this traffic doesn't stop soon I am going to get upset	ANGER
7	Listening to the new can be terrifying sometimes	FEAR
8	My car broke down today, I almost started crying. I am going to miss that car	SADNESS

Keep in mind that this is a small data set only for display purposes. The real data set we used for training is much larger than this. To show how our application works, let us use the following two sentences in Table 2 as target.

Table 2. Target sentences for emotion detection

SID	Target sentences	Emotion labels
1	This is so exciting! I am doing to Disney land!	(Unknown)
2	I was driving today and this person cut me off. I was so pissed off and upset	(Unknown)

The labels for these two sentences should end up being JOY and ANGER, respectively. The training file needs to have each sentence or document on its own line with the corresponding label separated by a tilde. This also means that there cannot be any tildes in the documents. The testing file should have each document on its own line as well.

When two files are supplied to our application, the process will look as follows. The application would take the training file and work with it first. In our example, it would take the training sentences and separate the documents from the labels. It would then take the documents and begin to preprocessing them. This would mean that it would stem all the words and remove any words that are in the stop words set. The outcome after the preprocessing is in Table 3.

Table 3. Training sentences and their emotion labels (after removing stop words)

SID	Training sentences (after removing stop words)	Emotion labels
1	Happi excit, won lotteri !	JOY
2	Today good day	JOY
3	Went shop today groceri store busi. people annoy	ANGER
4	Ahh terrifi thunderstorm	FEAR
5	Rain realli depress hope stop soon	SADNESS
6	Traffic doesn't stop soon upset	ANGER
7	Listen ne terrifi sometime	FEAR
8	Car broke today, start. miss car	SADNESS

After this, these documents would be transformed into a TF-IDF matrix. This sparse matrix would have dimensions 8×35. After this conversion, the TF-IDF matrix would be transformed into a smaller matrix. In our case, with this example, the matrix is already small, but we will run it through the SVD process for illustration purposes. We will choose the final size to be 8. The size of the final SVD matrix cannot be larger than the number of documents and must be smaller than the number of words. This will result in a dense matrix that is 8×8. This new smaller matrix is now used to train the SVM, while the TF-IDF matrix is used for the Naive Bayes classifier. The lexical classifier is not trained as its labels are based on the word value pairs that it is supplied within its lexicon. After the classifiers are trained, then the testing data is put through the same steps. The difference is that the documents are appended to the TF-IDF matrix. This means that new words are ignored, and the only words that are considered are words that were in the training data. The stemming and removal of the stops word from the testing sentences would result in Table 4.

Table 4. Target sentences for emotion detection (after removing stop words)

SID	Target sentences (after removing stop words)	Emotion labels
1	Excit ! disney land !	(Unknown)
2	Drive today person cut . piss upset	(Unknown)

Putting these through the TF-IDF matrix that was created from the training data results in a sparse matrix with only 6 values. These are at the words "excit", and "!" for the first sentence and at "upset", "today", and ",". Putting this matrix thought the same SVD process as the training data results in a dense 2×8 matrix. This matrix is then given to the SVD, while the TF-IDF matrix is given to the Naive Bayes classifier. The resulting labels are "JOY" and "ANGER" from both classifiers. The lexical classifier results in the same labels, as well as the first sentence, ends with values of 0.172 for anger, 0.132 for fear, 0.0 for sadness, and 0,697 for joy. For the second sentence, the lexical classifier gets the values 1.818 for anger, 0.484 for fear, 0.984 for sadness, and 0.0 for joy. Taking the maximum values returns the final labels of "JOY" and "ANGER" (Table 5).

Table 5. Target sentences and the analyzed emotion label

SID	Target sentences	Emotion labels
1	Excit ! disney land !	JOY
2	Drive today person cut. piss upset	ANGER

4 Evaluation and Analysis

The data set that we chose to use to train and test our classifier is made available by [9]. It has 3173 annotated tweets in it. For testing and training purposes, we spilled the tweets 70:30 giving us 2219 tweets for training and 954 for testing. The results that we obtained from this data set when classifying all emotions are seen in Table 6. The results for the binary classification of fear, joy, anger, and sadness can be seen in Tables 7, 8, 9, 10, respectively.

Table 6. Results of multi class labeling

	Precision	Recall	f1-score
Fear	0.84	0.77	0.80
Joy	0.84	0.75	0.80
Anger	0.48	0.94	0.64
Sadness	0.89	0.59	0.71

Table 7. Fear

	Precision	Recall	f1-score
Not fear	0.73	0.98	0.84
Fear	0.95	0.52	0.67

Table 8. Joy

	Precision	Recall	f1-score
Not joy	0.89	0.98	0.93
Joy	0.91	0.58	0.70

Table 9. Anger

	Precision	Recall	f1-score
Not Anger	0.99	0.97	0.98
Anger	0.68	0.80	0.74

Table 10. Sadness

	Precision	Recall	f1-score
Not sadness	0.81	98	0.89
Sadness	0.89	0.40	0.55

The low recall from the binary classifiers could be caused by a relative imbalance of labelled data. In those classifications, there is a larger volume of labelled data that falls under the "not" class when compared to the target class. However, the performance in the multi-label case is adequate overall. The notable exception of this is with the anger label. The drastically worse performance of this label warrants further research and investigation. Overall, this application gives us a very positive result in terms of emotion detection in Twitter text data.

5 Conclusion

As the textual form data is everywhere in our life nowadays, the demand for analyzing textual data is getting a lot of attention. One of the most essential types of textual analysis is sentiment analysis, which can detect/determine emotions from textual data. Furthermore, researchers today are no longer satisfied with simple positive/negative emotion detection anymore. We need applications that can detect more specific emotions like fear, joy, sadness, etc. In this paper, we present our work, which is an application that can analyze text data and determine the emotion of the text data. The evaluation result shows the practicality of our proposed application.

References

1. Bo, P., Lillian, L.: Seeing stars: exploiting class relationships for sentiment categorization with respect to rating scales. In: The 43rd Annual Meeting of the Association for Computational Linguistics, pp. 115–124. Association for Computational Linguistics, Ann Arbor (2005)
2. Cortes, C., Vapnik, V.: Support-vector networks. Mach. Learn. **20**(3), 273–297 (1995)
3. Da Silva, N., Hruschka, E., Hruschka Jr., E.: Tweet sentiment analysis with classifier ensembles. Decis. Support Syst. **66**, 170–179 (2014)
4. Hassan, S., Yulan, H., Miriam, F., Harith, A.: Contextual semantics for sentiment analysis of Twitter. Inf. Process. Manag. **52**(1), 5–19 (2016)
5. Huq, R., Ali, A., Rahman, A.: Sentiment analysis on Twitter data using KNN and SVM. Int. J. Adv. Comput. Sci. Appl. **8**(6), 19–25 (2017)
6. Ingo, F., Kurt, H.: wordnet: WordNet interface. R package version 0.1-14. https://CRAN.R-project.org/package=wordnet. Accessed May 2020
7. Loper, E., Bird, S.: NLTK: the natural language toolkit. In: The ACL Interactive Poster and Demonstration Sessions, pp. 214–217. Association for Computational Linguistics, Barcelona (2004)
8. Milgram, J., Cheriet, M., Sabourin, R.: "One against one" or "one against all": which one is better for handwriting recognition with SVMs? In: 10th International Workshop on Frontiers in Handwriting Recognition. Suvisoft, La Baule (2006)
9. Mohammad, S., Kiritchenko, S.: Understanding emotions: a dataset of tweets to study interactions between affect categories. In: The 11th International Conference on Language Resources and Evaluation. ELRA, Miyazaki (2018)
10. Oscar, A., Ignacio, C., Sánchez-Rada, J.F., Carlos, I.: Enhancing deep learning sentiment analysis with ensemble techniques in social applications. Expert Syst. Appl. **77**, 236–246 (2017)
11. Pak, A., Paroubek, P.: Twitter as a corpus for sentiment analysis and opinion mining. In: The 7th International Conference on Language Resources and Evaluation, pp. 1320–1326. ELRA, Valletta (2010)
12. Pang, B., Lee, L., Vaithyanathan, S.: Thumbs up?: sentiment classification using machine learning techniques. In: The ACL-2002 Conference on Empirical Methods in Natural Language Processing, vol. 10, pp. 79–86. Association for Computational Linguistics, Philadelphia (2002)

13. Pedregosa, F., Varoquaux, G., Gramfort, A., Michel, V., Thirion, B., Grisel, O., Blondel, M., Prettenhofer, P., Weiss, R., Dubourg, V., Vanderplas, J., Passos, A., Cournapeau, D., Brucher, M., Perrot, M., Duchesnay, E.: Scikit-learn: machine learning in Python. J. Mach. Learn. Res. **12**, 2825–2830 (2011)
14. Peter, D.-T.: Thumbs up or thumbs down? Semantic orientation applied to unsupervised classification of reviews (2002)
15. Rennie, J., Shih, L., Teevan, J., Karger, D.: Tackling the poor assumptions of naive Bayes text classifiers. In: 20th International Conference on Machine Learning, pp. 616–623. AAAI Press, Washington, DC (2003)
16. Schütze, H., Manning, C.-D., Raghavan, P.: Introduction to Information Retrieval. Cambridge University Press, Cambridge (2008)
17. Silva, C., Ribeiro, B.: The importance of stop word removal on recall values in text categorization. In: The International Joint Conference on Neural Networks, vol. 3, pp. 1661–1666. IEEE, Portland (2003)
18. Wiebe, J., Wilson, T., Cardie, C.: Annotating expressions of opinions and emotions in language. Lang. Resour. Eval. **39**(2–3), 165–210 (2005)
19. Zhang, L., Wang, S., Liu, B.: Deep learning for sentiment analysis: a survey. WIREs Data Min. Knowl. Discov. **8**(4), e1253 (2018)

Factors Affecting an Organization's Information Security Performance: The Characteristics of Information Security Officers

Ha-Kyeong Oh[1] and Tae-Sung Kim[2(✉)]

[1] Department of Information Security Management, Chungbuk National University, Cheongju, South Korea
gkrudlala@gmail.com

[2] Department of Management Information Systems, Chungbuk National University, Cheongju, South Korea
kimts@cbnu.ac.kr

Abstract. Information security breaches have damaged corporations' images and caused share prices to fall, and breaches of information security are emerging as an organizational risk. Various laws mandate that businesses employ information security officers with specialized knowledge, such a CPO, CIO, CSO, CISO, CPO, etc. Information security breaches affect general citizens as customers. However, despite the growing importance of information security officers, there is a lack of research on their role and characteristics. The purpose of this study is to analyze the roles of information security officers in various legal systems and the impact of the characteristics of information security officers on an organization's information security performance. A multiple linear regression is performed to validate the research model. The results of this study can act as a practical criterion for selecting information security officers who meet an organization's information security objectives.

Keywords: Information security officers · Information security performance · Individual characteristics

1 Introduction

According to the 2018 Veritas Global Data Privacy Consumer Study [1] report, customers are increasingly concerned about companies collecting personal information. That is why customers are willing to make a disadvantage for companies that do not properly protect customers' personal information. The personal information that a company holds is not only the property of the enterprise but also the private property of the information provider. Therefore, if any personal information leaks for any reason, it causes psychological and economic harm to the individual. And it causes corporate economic loss due to the accident response cost, victim compensation cost, and decline of the company image [2]. Customers' trust in public companies that experience information security incidents decreases, which may lead to a drop in their share price. In

W. Lee et al. (Eds.): *Big Data Analyses, Services, and Smart Data*, AISC 899, pp. 77–84, 2021.
https://doi.org/10.1007/978-981-15-8731-3_6

April 2016, Facebook's share price fell more than 16% after details of the company's user information case was released. In July 2018, Equifax, one of the three major credit rating agencies in the United States, suffered a sharp decline in share price due to the leakage of personal credit information.

As the variety of cyberattacks is increasing, organizations should have a variety of information security officers with the expertise to respond to information security incidents. According to the Personal Information Protection Act, the Act on Promotion of Information and Communications Network Utilization and Information Protection, the Framework Act on National Informatization, the Act on the Protection of Information and Communications Infrastructure, and the General Data Protection Regulation (GDPR), organizations must appoint a person in charge of information security. However, despite the growing importance of information security officers, there is a lack of research on their characteristics.

The purpose of this study is to investigate the impact of the characteristics of information security officers on an organization's information security performance. The results of this study are expected to be a practical criterion for selecting and empowering information security officers who meet the organization's information security goals.

2 Literature Review

Wee et al. [3] classify characteristics as roles, personal traits, and background knowledge, and explain the effects of the characteristics on an organization's personal information security performance. Therefore, in this study, the characteristics of information security officers are classified as roles, personality traits, and background knowledge. The correlation between these characteristics and organizational performance was examined and is summarized in Table 1.

Table 1. Literature review of characteristics and organizational performance.

Classification	Study
The role of information security officer	[3, 4]
Background knowledge of information security Officer	[5, 6]
Characteristics and organizational performance of the management	[7–10]

3 Research Design

3.1 Research Model

In this study, we derived a research model assuming that the characteristics of the officers in charge of information security affect an organization's information security performance (Fig. 1, Research Model). The results of the literature review were synthesized, and the factors that influence the characteristics are suggested from two perspectives. Characteristics were classified as roles and background knowledge. The role of an information security officer is divided into the following roles: an information providing role, a strategic role, and an external relations role. Background knowledge is divided into three types: legal, technical, and management.

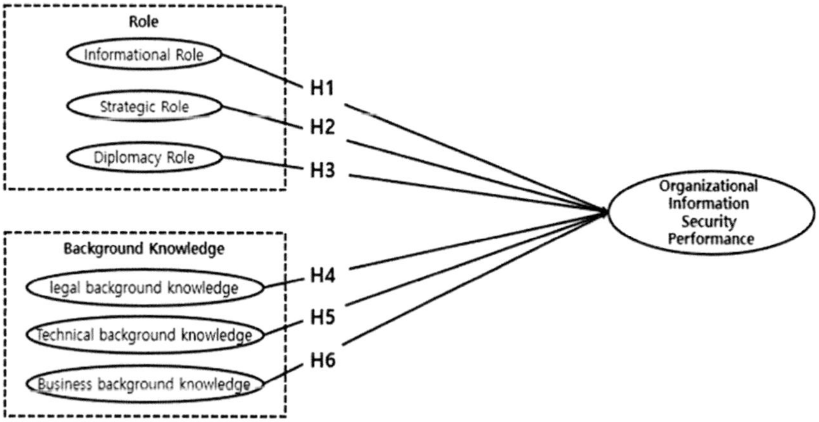

Fig. 1. Research model

3.2 Hypotheses

The hypotheses of this study were established based on the research model designed based on the existing research.

First, the role of information security officers has a significant impact on the organization's information security performance. The following hypotheses are suggested:

H1: The informational role has a positive impact on the organization's information security performance.
H2: The strategic role has a positive impact on the organization's information security performance.
H3: The diplomacy role has a positive impact on the organization's information security performance.
Second, the background knowledge of information security officers has a significant impact on the organization's information security performance. The following hypotheses are established:
H4: Legal background knowledge has a positive impact on the organization's information security performance.
H5: Technical background knowledge has a positive impact on the organization's information security performance.
H6: Business background knowledge has a positive impact on the organization's information security performance.

4 Research Methods and Results

4.1 Study Participants

In this study, we collected data through a questionnaire survey distributed to information security executives online and offline from August 30, 2018, to November 24, 2018. A total of 133 questionnaires were collected, and 101 questionnaires, excluding those with incomplete or inconsistent responses, were included in the final analysis. Positions were divided into Chief Privacy Officer (CPO), Chief Information Officer (CIO), Chief Security Officer (CSO), Chief Information Security Officer (CISO), and Data Protection Officer (DPO). The demographic characteristics of the respondents are shown in Table 2.

Table 2. Demographic characteristics.

Category	Subcategory	# of Respondents(%)
Type of organization	Small-sized business	9(8.9%)
	Medium-sized business	21(20.8%)
	Large business	25(24.8%)
	Public organization/Government	30(29.7%)
	Venture	6(5.9%)
	Others	10(9.9%)
Industry	Manufacturing/distribution	8(7.9%)
	Construction	6(5.9%)
	Networking/information services	37(36.6%)
	Financial services/insurance	18(17.8%)
	Science and technical service	14(13.9%)
	Business facilities management/business support services	5(5.0%)
	Others	13(12.9%)
Job title	Chief Privacy Officer (CPO)	31(30.7%)
	Chief Information Officer (CIO)	4(4.0%)
	Chief Security Officer (CSO)	34(33.7%)
	Chief Information Security Officer (CISO)	30(29.7%)
	Data Protection Officer (DPO)	2(2.0%)

In the questionnaire, the type of organization was divided into small-sized business, medium-sized business, large business, public organization/government and venture. Industries include manufacturing/distribution, construction, networking/information services, financial services/insurance, science and technical service and business facilities management/business support services.

The characteristics of the sample groups selected in this study are as follows. Information security officers are widely distributed in the networking/information service industry and the financial service/insurance industry. This is because the law mandates companies employ information security officers. To securely manage information systems, as well as human and material security systems, at the enterprise level, companies increasingly are appointing CSOs, the senior executives responsible for managing and controlling information security. In addition, as awareness of the need for personal information privacy and concern about personal information infringement increase, as well as the issue of personal information leakage, the number of companies appointing CPOs is increasing [11]. As the Act on the Protection of Information and Communications Infrastructure, which stipulates the qualifications of CISO, has been implemented, more and more companies are appointing CISOs. Thus, the number of CIO samples is considered to be small. In May 2018, the GDPR, a law that is binding for all European Union (EU) members, was implemented. The GDPR applies to companies in the EU and all countries that do business with these companies. With the implementation of the GDPR, enterprises subject to the application of the GDPR must appoint a DPO. Few companies have appointed DPOs because of the recent GDPR implementation. Therefore, the sample was insufficient.

4.2 Research Method

The data collected through the survey were analyzed using SPSS Statistic 22.0. The research model is a multiple linear regression model with six independent variables, informational role, strategic role, diplomacy role, legal background knowledge, technical background knowledge and business background knowledge, to explain changes in the organization's information security performance. Multiple linear regression analysis was performed to determine which independent variables influenced each other in a complex relationship. The multiple linear regression equation can explain that the linear sum of each datum is ultimately related to the dependent variable. This is based on the basic regression equation shown in Eq. (1). Based on this equation, the regression equation was derived. Y_i is the organization's information security performance, X_1 is the informational role, X_2 is the strategic role, X_3 is the diplomacy role, X_4 is legal background knowledge, X_5 is technical background knowledge, and X_6 is management background knowledge. i is 1, 2, ..., n.

$$Y_i = \beta_0 + \beta_1 X_{1i} + \beta_2 X_{2i} + \beta_3 X_{3i} + \beta_4 X_{4i} + \beta_5 X_{5i} + \beta_6 X_{6i} + \varepsilon_i. \quad (1)$$

4.3 Results

The existence of a multiple linear regression model between an independent variable and a dependent variable was verified with analysis of variance ($p = 0.000 < 0.5$). Table 3 shows the results of the regression analysis to verify the impact of the characteristics of the information security officer on the organization's information security performance.

Table 3. The relationship between characteristics and an organization's information security performance.

Independent variable	Hypothesis	t-value	p-value	Tolerance
(Constant)		1.727	0.088	
Informational role	H1	2.367	0.020	0.748
Strategic role	H2	−1.009	0.315	0.312
Diplomacy role	H3	1.790	0.077	0.393
Legal background knowledge	H4	4.877	0.000	0.705
Technical background knowledge	H5	2.241	0.027	0.560
Business background knowledge	H6	0.070	0.945	0.389
Dependent variable: information security performance				

The dependent variables showed a high correlation (0.711), and the explanatory rate explaining the independent variables was 50.5%. The predictive model of this study through the regression equation of Eq. (1) is as shown in Eq. (2):

$$Y_i = 1.727 + 2.367X_{1i} - 1.009X_{2i} + 1.790X_{3i} + 4.877X_{4i} + 2.241X_{5i} + 2.367X_{6i}. \tag{2}$$

Assuming that the independent variable and the dependent variable can be expressed numerically, the organization's information security performance increases by 2.367 when the informational role increases by 1. As the strategic role increases by 1, the organization's information security performance decreases by 1.009. If the diplomacy role increases by 1, the organization's information security performance increases by 1.790. If legal background knowledge increases by 1, the organization's information security performance increases by 4.877. If the technical background knowledge increases by 1, the organization's information security performance increases by 2.241. If business background knowledge increases by 1, the organization's information security performance increases by 2.367. The hypothesis that informational role, legal background knowledge, and technical background knowledge positively affect an organization's information security performance has a t-value greater than 1.96. H1, H4, and H5 were supported and statistically significant at a p value less than 0.05. However, strategic role, diplomacy role, and business background knowledge had a t-value of less than 1.96 and a p value greater than 0.05. The regression coefficients for strategic role, diplomacy role, and business background knowledge were not statistically significant. Therefore, H2, H3, and H6 were not supported. The hypothesis verification results are shown in Fig. 2.

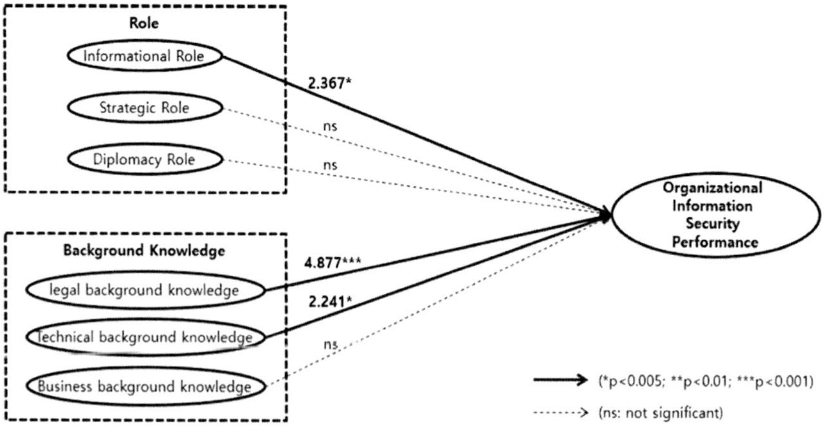

Fig. 2. Analysis results

5 Conclusion

In this study, we hypothesized that the characteristics of the information security officer are correlated with the company's information security performance. We surveyed information security officers, collected relevant data, and analyzed them empirically. We used SPSS Statistics 22.0 to analyze the data; the measurement models and structural models based on responses from the survey.

The roles of information security officers who affect their organization's information security performance are informational, strategic, diplomacy, and the background knowledge is legal, technical, and business.

Based on the hypothesis testing, the informational role of information security officers has a statistically significant influence on an organization's information security performance. The role of providing information is considered basic, and is the duty of the information security officer. The strategic and diplomacy roles did not have a statistically significant impact on the organization's information security performance.

Regarding background knowledge held by the information security officer, legal and technical background knowledge was shown to have a positive effect on the performance of the organization's information security. It seems that the importance of responding to the law in information security affairs affected the correlation between the background knowledge and the information security performance. In South Korea, information security officers often have technical background knowledge, and it is necessary to have technical background knowledge to perform information security work. However, business background knowledge did not have a statistically significant effect on an organization's information security performance.

The value of information assets has increased, and the importance of information security has increased in the business environment. Businesses have pursued a high level of security. As a result, businesses are aware of the need for personnel to protect information and supervise security management systems. The result of this study will be an important criterion in hiring information security executives and empowering them.

In addition, information security officers will be able to judge the competencies needed for career development related to these jobs.

Acknowledgements. This work was supported by the Ministry of Education of the Republic of Korea and the National Research Foundation of Korea (NRF-2018S1A5A2A01039356).

References

1. Veritas: Veritas Global Data Privacy Consumer Study (2018)
2. Yoo, J.H., Jie, S.H., Lim, J.I.: Estimating direct costs of enterprises by personal information security breaches. J. Korea Inst. Inf. Secur. Cryptol. **19**(4), 63–75 (2009)
3. Wee, J.Y., Jang, J.Y., Kim, B.S.: The impact of CPO characteristics on organizational privacy performance. Asia Pac. J. Inf. Syst. **24**(1), 95–114 (2014)
4. Mintzberg, H.: Managerial work: analysis from observation. Manag. Sci. **18**(2), B–97 (1971)
5. Applegate, L.M., Elam, J.J.: New information systems leaders: a changing role in a changing world. MIS Q. **16**(4), 469–490 (1992)
6. Bennett, S.C.: Do you need a chief privacy officer. Pract. Lawyer **53**, 17–20 (2007)
7. Hambrick, D.C., Mason, P.A.: Upper echelons: the organization as a reflection of its top managers. Acad. Manag. Rev. **9**(2), 193–206 (1984)
8. Hart, S.L., Quinn, R.E.: Roles executives play: CEOs, behavioral complexity, and firm performance. Hum. Relat. **46**(5), 543–574 (1993)
9. Banker, R.D., Hu, N., Pavlou, P.A., Luftman, J.: CIO reporting structure, strategic positioning, and firm performance. MIS Q. **35**(2), 487–504 (2011)
10. Li, Y., Tan, C.H.: Matching business strategy and CIO characteristics: the impact on organizational performance. J. Bus. Res. **66**(2), 248–259 (2013)
11. Korea Internet & Security Agency: 2012 Survey on the Information Security (2012)

An Empirical Investigation of Customer Loyalty in Chinese Smartphone Markets with Large-Scale Data: Apple, Samsung, and Xiaomi Cases

Hyung Jin Kim[1](✉), Xiao Qing Ding[2], and Ho Geun Lee[2]

[1] Korea Institute of Industrial Technology, 143 Hanggaulro, Sangrok-gu, Ansan-si, Gyeonggi-do 15588, South Korea
kimhyungjin@kitech.re.kr

[2] School of Business, Yonsei University, 50 Yonsei-ro Seodaemun-gu, Seoul 03722, South Korea
fairybaby@live.com, h.lee@yonsei.ac.kr

Abstract. In recent years, the eco-systems consisting of hardware, operating systems, applications and other services have increasingly become important for mobile phone businesses, and the diffusion of smartphones expedites this trend at a rapid pace. Moreover, the shortened lifespan of smartphones makes it harder for smartphone companies to retain their customers for a long time. In this study, we aim to identify the antecedents of smartphone users' satisfaction and loyalty (CS&L) and then examine how the three primary components of smartphones, i.e., hardware, software, and service, can affect CS&L differently through their cognitive and affective beliefs. We developed a theoretical model based on the cognitive-affective system theory of personality, and then tested the research model with large-scale empirical data (N = 710), collected from Chinese smartphone users. In an additional analysis, we compared the leading smartphone brands in China (i.e., Apple, Samsung, and Xiaomi) in terms of customer switching, and found that different strategic focuses of those companies on hardware, software, and service have dissimilar impacts on CS&L.

Keywords: Smartphone hardware · Smartphone software · Smartphone service · Relative advantage · Emotional attachment · Customer satisfaction · Customer loyalty

1 Introduction

One decade ago, the competition of conventional mobile phones primarily focused on the quality of hardware. In recent days, however, mobile eco-systems consisting of hardware, mobile operating systems, application stores, cloud services, etc. have increasingly become important for the competitiveness [25]. Indeed, the emergence and diffusion of smartphones expedites this trend at a rapid pace.

This article is revision of the second author's master's thesis from Yonsei University

W. Lee et al. (Eds.): *Big Data Analyses, Services, and Smart Data*, AISC 899, pp. 85–105, 2021.
https://doi.org/10.1007/978-981-15-8731-3_7

As a more computing power is required for processing heavier data more quickly than before, the quality of hardware is still important for smartphones. However, hardware may not be as dominant as for conventional mobile phones. The operating systems (OS) and other software such as mobile apps run on the systems are now considered another primary component of smartphones. Further, because smartphones are much more complicated in use than conventional mobile phones, customers may frequently encounter troubles and needs when they use. Therefore, the service provided by smartphone companies may also influence customer usage experience significantly. In addition, the lifespan of smartphones is quite short, which makes it harder for smartphone companies to retain their customers for a long time. It is thus a crucial question for the management how to map out a strategy for satisfying customers and having them loyal to their brands.

In this paper, we attempt to extend the extant customer satisfaction and loyalty (CS&L) literature to the context of smartphones. Our study aims to discover the antecedents of smartphone users' CS&L and examine how the three main components of smartphones, i.e., hardware, software, and service, can affect CS&L differently. We developed a theoretical model based on the cognitive-affective system theory of personality, and then tested the research model with large-scale empirical data collected from Chinese smartphone users.

Chinese mobile phone industry has grown rapidly, and the number of users already exceeded 1.3 billion in 2015 [57]. And, the two leading brands, i.e., Samsung and Apple, compete with Chinese local brands such as Xiaomi and Huawei. So, another interesting research question is how the top two brands differ from a local brand, i.e., Xiaomi, in terms of their CS&L strategies for the Chinese mobile market.

Our data analysis using large-scale data (N = 710) shows that all of our hypotheses were supported, suggesting that smartphone users' CS&L can be explained with both cognitive and affective beliefs (i.e., relative advantage, emotional attachment), and that hardware performance, software quality, and service quality can directly determine the degree of those beliefs. Another interesting finding is that in the Chinese mobile market, Samsung, Apple, and Xiaomi show different strategic focuses on the three competing factors for smartphones (i.e., hardware, software, and service).

Our study contributes to both CS&L literature and smartphone literature. Specifically, our results provide evidence that smartphone users' CS&L can be explained from the perspective of cognitive-affective system theory of personality (CAPS). Our model suggests that cognitive beliefs and affective beliefs in CAPS can be specifically measured with the perceived relative advantage variable and the perceived emotional attachment variable, respectively. Our study also finds that the cognitive and affective beliefs of smartphone users can be formed by smartphone companies' strategic differences on hardware performance, software quality, and service quality. Moreover, our specific factors for the three categories of mobile strategy provide practitioners with an opportunity to evaluate and further improve their current strategies for CS&L.

2 Literature Review

2.1 Customer Satisfaction and Loyalty (CS&L)

There is a large body of literature studying customer satisfaction and customer loyalty about mobile phones and telecommunications. Indeed, earlier studies on CS&L date back to years ago. One of the research topics that prior research has paid much attention to is finding the antecedents of customer loyalty. For example, Aydin and Özer [1] analyzed the antecedents of customer loyalty in the mobile telecommunication market. In that model, perceived service quality, trust, corporate image, and switching cost are considered as the factors affecting customer loyalty. More recently, Lee [31] studied customers' loyalty to smartphones and mobile communications service providers. By analyzing the impacts of corporate image, price, product quality, and product innovation, she found that corporate image and product quality are the key drivers for customer loyalty.

Customer satisfaction has also been studied in many different contexts. To explore its determinants, researchers have attempted to consider several factors. Chang et al. [11] investigated the determinants of customer satisfaction and continuance intention of smartphones by proposing five factors, i.e., network quality, system quality, contents quality, customer support, and compatibility. Gijón et al. [18] examined a model of customer satisfaction of mobile phones that consists of three antecedents of customer satisfaction, i.e., perceived value, perceived quality, and image of the carrier. There has also been research effort to identify more specific factors that lead to customer satisfaction. For instance, Haverila [20] discussed the relationship among mobile phone preferences, customer satisfaction, and repurchase intent of customers, and then found that battery/talk time was the most important feature of mobile phones leading to the positive effect.

Customer loyalty has often been studied together with customer satisfaction in the literature. Shin [51] discussed how customer usage experience affects customer satisfaction of smartphones and showed that perceived value and customer satisfaction of smartphones were the key factors that mediated the relationship between quality and customer loyalty. Liang et al. [33] discussed the relationship between service quality and customer switching (i.e., not loyal) in China's mobile phone service market. Their results explain that Chinese mobile phone users have a tendency to switch to a new service provider even when the current service is satisfactory.

Across the literature, CS&L has received much attention from researchers, especially with regard to what leads to CS&L. However, from the managerial point of view of smartphone companies, how to encourage CS&L is still an ongoing question to sustain their businesses in a particular market. This is due in part to the inability of prior studies to give a theoretical and practical idea for evaluating and improving CS&L, based on the contemporary trend of the market. Especially, the recently-found competing factors in smartphone markets, i.e., hardware performance, software quality, and service quality, have little been considered as the very-first factors exerting the influence on CS&L and relating to smartphone companies' strategies.

3 Theoretical Development

Our study builds a theoretical model that explains how customer loyalty arises in the context of smartphone usage. From the theoretical viewpoint, our research model is based on the cognitive-affective system theory of personality (CAPS), which focuses on the two contrasting but complementing aspects, i.e., cognitive beliefs and affective belief [16].

Mischel and Shoda [37] proposed the CAPS to reconcile paradoxical findings on the variability of behavior and the invariance of personality. In the theory, individuals in a situation are assumed to differ in the accessibility of cognitive-affective mediating units, such as affects and beliefs. The features of a situation are assumed to be able to activate a set of internal reactions based on an individual's prior experience [38]. The reactions can be both cognitive and affective. Then, all of the activated mediating units, i.e., beliefs and affects, may generate behaviors of the individual, which is described as a behavior generation process [e.g., 37, 38].

Many prior studies have relied on the theory to examine user reactions to IT products and services. Especially, previous studies using the theory have investigated users' willingness to remain with a current product, suggesting that the theory is effective to explain how a long-term orientation of users is generated in both cognitive and affective ways. For example, several researchers have found that both cognitive and affective factors influence intentions to continuously adopt online services [e.g., 30, 45] and to continuously use agile information systems [21]. Further, the theory corresponds to an existing perspective on the satisfaction process [35]; that is, cognitive and affective elements are antecedents of satisfaction, which in turn lead to customer loyalty.

In our research model, we conceptualize the two beliefs that determine further behavior of smartphone users, in terms of perceived relative advantage and perceived emotional attachment, respectively. Then, we attempt to identify specific features of smartphones that lead to the cognitive and/or affective beliefs. Prior studies have shown that perceived relative advantage and perceived emotional attachment are positively associated with CS&L [e.g., 9, 15, 34, 42, 43].

To explore the specific features of smartphones affecting users' cognitive and affective beliefs, we focus on the three competing elements of smartphones: hardware performance, software quality, and service quality. The extant literature in CS&L of conventional mobile phones has found that hardware quality is one of the important factors for CS&L [e.g., 23, 24, 36]. Given that smartphones have more complicated hardware structure and components for better performance than before, hardware quality can still remain a key driver for users' CS&L. Other previous studies suggest that mobile OS, applications, and after-sales services are also playing increasingly important roles in satisfying diverse user needs in these days [23]. Based on this literature review, we develop a research model shown in Fig. 1.

3.1 The Effect of Customer Satisfaction on Customer Loyalty

Customer loyalty refers to a long-term commitment to repurchasing, which involves both repeated patronage and favorable attitude towards a product or service [26]. Prior studies have suggested that the overall assessment of a product can affect one's willingness to remain with the same product, i.e., loyalty [32]. In the literature, customer satisfaction is typically defined as an overall assessment of the performance of various attributes that constitute a product or service [e.g., 17, 22, 50, 54]. Hence, we anticipate that:

Hypothesis 1 (H1). Customer satisfaction with a smartphone has a positive effect on customer loyalty to the smartphone.

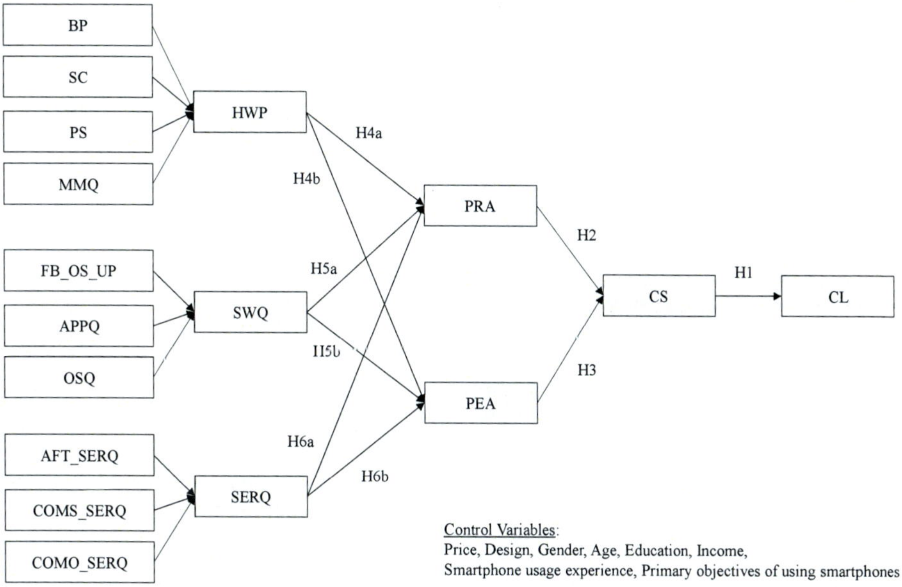

Fig. 1. Research model. *Note*: BP = battery performance, SC = storage capacity, PS = processor speed, MMQ = multi-media quality, FB_OS_UP = feedback based OS update, APPQ = application quality, OSQ = OS quality, AFT_SERQ = after-sale service quality, COMS_SERQ = community support service quality, COMO_SERQ = communication opening service quality, HWP = hardware performance, SWQ = software quality, SERQ = service quality, PRA = perceived relative advantage, PEA = perceived emotional attachment, CS = customer satisfaction, CL = customer loyalty

3.2 The Effect of Perceived Relative Advantage on Customer Satisfaction

Relative advantage refers to the degree to which using an innovation is perceived as being better than its predecessor [48]. It is usually perceived when we have experienced more than one product and compare those products with each other [47]. According to prior studies, relative advantage is positively associated with acceptance of a product [13], and other research indicates that user acceptance of a product is affected by satisfaction levels with the product. Accordingly, satisfaction levels may have a mediating relationship between relative advantage and acceptance. Therefore, we hypothesize that:

Hypothesis 2 (H2). Perceived relative advantage of a smartphone has a positive effect on customer satisfaction with the smartphone.

3.3 The Effect of Perceived Emotional Attachment on Customer Satisfaction

According to Bowlby [8], an attachment is defined as an emotion-laden target-specific bond between a specific object and an individual. Some products can inspire an internal connection with customers over time [55]. Prior research has shown that high attachment levels can influence a consumer' positive opinion on the value of the product [55]. This is because a person who is emotionally attached to a brand or product is likely to find it satisfactory [18]. In this regard, we believe that perceived emotional attachment to a smartphone that can serve as a user's affective evaluation that can positively affect his or her satisfaction with the smartphone.

Hypothesis 3 (H3). Perceived emotional attachment to a smartphone has a positive effect on customer satisfaction with the smartphone.

3.4 The Effect of Hardware Performance on Perceived Relative Advantage and Perceived Emotional Attachment

Churchill and Surprenant [14] defines performance as how well a product works in terms of intended purposes. Smartphones require a higher level of hardware performance than conventional mobile phones, due to the various functions included. Previous research has studied the influence of specific features of smartphones on CS&L, such as battery performance, storage capacity, camera quality [e.g., 23, 24, 36]. As smartphones are used whenever necessary in daily life, the *battery performance* or capacity is a critical component of hardware performance [7]. In addition, a great number of users tend to send and save heavy files on their smartphones, such as electronic documents, photos, and movies. Accordingly, *storage capacity* can also be a measurement of smartphone hardware performance [e.g., 24, 53]. *Multi-media quality* and *processor spee*d are another set of features that determines the hardware performance of smartphones [e.g., 2, 23]. Nowadays, people take pictures, listen to music, and watch movies with their high performance smartphones. In light of this, we expect that the hardware performance of a smartphone can affect users' cognitive evaluations and emotional attachment to their smartphones.

Hypothesis 4A (H4A). Hardware performance of a smartphone has a positive effect on perceived relative advantage of the smartphone.

Hypothesis 4B (H4B). Hardware performance of a smartphone has a positive effect on perceived emotional attachment to the smartphone.

3.5 The Effect of Software Quality on Perceived Relative Advantage and Perceived Emotional Attachment

In recent years, software-related smartphone markets are getting bigger and bigger [27]. Mobile apps run on smartphones are closely related to almost every part of our daily life, such as work schedule, weather forecast, shopping, game, and so on. Thus, *the quality of mobile applications* has become a unique feature of smartphones that affect users' evaluation [28]. Next, another significant software element for smartphones is *operating systems* (OS). The release of mobile OS updates and the quality of a recent OS version are associated with the robustness and expandability of smartphones, thereby influencing users' evaluations and emotions significantly [51]. Moreover, given that most valuable comments on smartphone OS can be provided by individual users, *user feedback-based OS updates* recently matter to users' diverse decisions on smartphones. Therefore, we hypothesize that:

Hypothesis 5A (H5A). Software quality of a smartphone has a positive effect on perceived relative advantage of the smartphone.

Hypothesis 5B (H5B). Software quality of a smartphone has a positive effect on perceived emotional attachment to the smartphone.

3.6 The Effect of Service Quality on Perceived Relative Advantage and Perceived Emotional Attachment

Service quality is another important factor in the customer loyalty literature [29]. Smartphones are expensive products and customers have high expectations at the time of purchase. Therefore, smartphone sellers need to provide satisfactory after-sale service support to customers in terms of the accessibility of service centers, the amount of service fees, and the service time [e.g., 5, 19]. *Community support* is another type of popular practice used by various companies to improve CS&L [56]. The high performance technical specifications and various functions of a smartphone can be a difficulty for those users who are not very familiar with technology. In this case, online user community site and service sponsored by the company can help those users effectively, likely affecting their cognitive and affective evaluations [4]. Lastly, customers are normally required to sign contracts with telecommunication companies for their smartphones to *open the service*. There often exist some cases wherein customers have to switch to a new service set or even a new telecommunication company in order to use a smartphone they want. Those processes are costly and the inconvenience may lead to negative reactions of users. So, we believe that:

Hypothesis 6A (H6A). Service quality of a smartphone has a positive effect on perceived relative advantage of the smartphone.

Hypothesis 6B (H6B). Service quality of a smartphone has a positive effect on perceived emotional attachment to the smartphone.

4 Data Analysis

4.1 Data Collection

Our model was tested with the data collected from smartphone users of three popular brands in China, i.e., Apple, Samsung, and Xiaomi. Chinese smartphone market is one the largest smartphone markets in the world; hence, our comprehensive model of CS&L can be tested in diversity of data, likely providing significant implications.

Data from 710 users, who were randomly selected, was collected in two ways. Firstly, online survey was conducted for three weeks in several Chinese online platforms. Totally, 390 responses were collected (147 Xiaomi' users, 145 Apple' users, and 98 Samsung's users). After the first survey, another data collection was done with a social investigation company for one week. Data from 100 Xiaomi' users, 100 Apple' users, and 120 Samsung's users were then collected. The respondents were told that their participation should be voluntary and they would be rewarded with 10 Chinese Yuan each (approximately 2 US dollars).

For the surveys, online survey systems were designed to record the time that respondents spent to complete the questionnaire and their IP addresses. The responses that were finished within an unusual short time or in the same IP address were finally dropped before data analysis.

4.2 Measurement

Our measurement items were mostly derived from validated scales in the relevant literature. Firstly, customer loyalty was measured with the items borrowed from prior studies (e.g., [58]). The measures of customer satisfaction were used based on the literature (e.g., [46]). For the perceived relative advantage, we used the items that were also from previous research (e.g., [40]). Similarly, the perceived emotional attachment was measured with existing items developed in the literature (e.g., [3]).

Next, our model includes three 2nd-order factors (i.e., hardware performance, software quality, and service quality). The first-order factors for hardware performance were measured based on the concepts defined in the literature. To measure battery performance, we used items asking whether the battery is long-lasting and it is charged quickly, for example [39]. Storage capacity was measured with the items such as whether the internal storage is sufficient and external memory extension is available [24]. Processor speed refers to how powerful and fast the CPU of a smartphone can process data [10]. The measures of multimedia quality focus on how the camera, video, and sound quality are good and satisfactory [23].

For software quality, the feedback based OS update was defined with the measures such as whether the company listens to customer voices or expectations for OS updates [Sperry 2014]. The definition of application quality is quite straightforward, meaning that smartphone apps are useful and satisfactory, and are provided in variety [52]. The OS quality measures include whether the mobile OS performs reliably and it recovers quickly from errors [41].

We used the items developed in [44] to measure After-sale service quality in terms of easy access, prompt responses, and friendliness. The measures of community support service quality were used based on [49]. Lastly, we developed the measures of communication opening service quality by focusing on whether the subscription process is not complicated, a multi-year contract is required, and the like. All of the questions were measured with 7 point Likert scales (1 = "strongly disagree," 4 = "Neutral," 7 = "strongly agree").

4.3 PLS for Data Analysis

Since the smartphone feature variables in our research model are conceptualized as second-order factor models, we use PLS (Partial Least Squares) method for data analysis. PLS is a structural equation modeling method of casual modeling which utilizes a component-based approach to estimation. In contrast to the covariance-based SEM methods such as LISREL and EQS which use a maximum likelihood function to obtain parameter estimates, the component-based method uses a least squares estimation procedure. PLS is known to allow flexibility for testing both formative and reflective latent constructs and place low requirement on sample size, measurement scales, and distributional assumptions [e.g., 12, 17].

5 Results

5.1 Descriptive Analysis

The total set of our data consists of 247 Xiaomi's users, 218 Samsung's users, and 245 Apple's users. There was no significant difference among the three different brands in terms of demographic statistics. Most of the respondents reported that they had switched to their current smartphones in the last one year. For each brand, more than 60% of users are within the age of 20 to 40, and more than 60% users have received university educations. Most of the users have the annual income between 50,000 and 10,000 Chinese Yuan, (approximately 10,000 to 20,000 US dollars), and they have experienced smartphones longer than one year. The respondents are found to use smartphones for entertainment, social communication, and information retrieval (Table 1).

Table 1. Descriptive statistics

Demographics	Total (N = 710)	Xiaomi N = 247	Samsung N = 218	Apple N = 245
		Frequency (%)	Frequency (%)	Frequency (%)
The duration of using the current smartphone	Less than 1 year	227(91.9)	186(85.3)	200(81.6)
	1 2 years	17(6.9)	27(12.4)	34(13.9)
	2–3 years	3(1.2)	4(1.8)	10(4.1)
	Equal or above 3 years	0(0.0)	1(0.5)	1(0.4)
Gender	Male	132(53.4)	110(50.5)	119(48.6)
	Female	115(46.6)	108(49.5)	126(51.4)
Age	Less than 20 years old	18(7.3)	10(4.6)	2(0.8)
	20–40 years old	175(70.9)	150(68.8)	204(83.3)
	More than 40 years old	54(21.9)	58(26.6)	39(15.9)
Education	Below high school	6(2.4)	10(4.6)	2(0.8)
	High school	22(8.9)	8(3.7)	5(2.0)
	Bachelor's degree	154(62.3)	148(67.9)	167(68.2)
	Master's degree	48(19.4)	38(17.4)	51(20.8)

(continued)

Table 1. (*continued*)

Demographics	Total (N = 710)	Xiaomi N = 247	Samsung N = 218	Apple N = 245
		Frequency (%)	Frequency (%)	Frequency (%)
	Above Master's degree	17(6.9)	14(6.4)	20(8.2)
Annul income (unit: thousand Chinese yuan)	Less than 50	54(21.9)	46(21.1)	60(24.5)
	50–100	95(38.5)	75(34.4)	84(34.3)
	100–150	61(24.7)	64(29.4)	61(24.9)
	150–200	22(8.9)	18(8.3)	19(7.8)
	Equal or Above 200	15(6.1)	15(6.9)	21(8.6)
Smartphone usage experience	Less than 1 year	7(2.8)	8(3.7)	6(2.4)
	1–2 years	32(13.0)	27(12.4)	21(8.6)
	2–3 years	62(25.1)	40(18.3)	42(17.1)
	3–4 years	58(23.5)	48(22.0)	70(28.6)
	4–5 years	56(22.7)	57(26.1)	55(22.4)
	Equal or above 5 years	32(13.0)	38(17.4)	51(20.8)
Primary objective(s) of using smartphones	Handling tasks	127(16.2)	115(16.6)	141(16.8)
	Amuse activities	177(22.5)	165(23.8)	213(25.4)
	Social activities	226(28.8)	187(26.9)	211(25.2)
	Education and learning activities	100(12.7)	88(12.7)	95(11.4)
	Information search	147(18.7)	130(18.7)	166(19.8)
	Etc.	8(1.0)	9(1.3)	11(1.3)

5.2 Test of Measurement Model

Before testing the hypothesized structural model, the measurement model was first assessed in terms of internal consistency reliability, and convergent and discriminant validity [e.g., 6, 12]. Since our model contains formative second-order constructs (e.g., HWP, SWQ, and SERQ), we divided the measurement model into two parts: *part1* for the 1st-order factors and *part2* for the 2nd-order factors and other constructs.

For *part 1*, one of the items measuring storage capacity (i.e., sc2) was omitted due to low loading. The item os4 was also dropped because of the high correlation between OS quality and application quality. Then, all of the remaining item loadings were greater than 0.707, and no items loaded higher on constructs they were not intended. In addition, the composite reliability scores range from 0.85 to 0.95, which indicates high internal consistency. The lowest AVE value, 0.664 exceeds the cut-off value of 0.50, indicating that the requirement for measurement reliability was met. Finally, discriminant validity was also confirmed, as the square roots of the AVE between constructs and their measures were all larger than the correlations between constructs in the same row and column (Tables 2 and 3).

Table 2. Convergent validity and internal consistency reliability for the 1st-order constructs

	Loadings	Weight	CR	AVE	α
BP	bp1	0.858	0.862	0.677	0.760
	bp2	0.857			
	bp3	0.749			
SC	sc1	0.931	0.92	0.852	0.826
	sc3	0.915			
PS	ps1	0.920	0.942	0.845	0.908
	ps2	0.910			
	ps3	0.928			
MMQ	mmq1	0.840	0.890	0.730	0.817
	mmq2	0.880			
	mmq3	0.844			
FB_OS_UP	fb_os_up1	0.901	0.929	0.813	0.885
	fb_os_up2	0.923			
	fb_os_up3	0.879			
APPQ	appq1	0.881	0.895	0.741	0.824
	appq2	0.880			
	appq3	0.819			
OSQ	osq1	0.823	0.856	0.664	0.748
	osq2	0.783			
	osq3	0.829			
AFT_SERQ	aft_serq1	0.827	0.891	0.732	0.817
	aft_serq2	0.885			
	aft_serq3	0.854			
COMS_SERQ	coms_serq1	0.879	0.915	0.783	0.861
	coms_serq2	0.904			
	coms_serq3	0.871			
COMO_SERQ	como_serq1	0.888	0.880	0.709	0.805
	como_serq2	0.855			
	como_serq3	0.781			

Table 3. Discriminant validity for the 1st-order constructs

	(1)	(2)	(3)	(4)	(5)	(6)	(7)	(8)	(9)	(10)
(1) AFT_SERQ	**0.856**									
(2) APPQ	0.377	**0.861**								
(3) BP	0.353	0.35	**0.823**							
(4) COMO_SERQ	0.321	0.327	0.27	**0.842**						
(5) COMS_SERQ	0.502	0.428	0.276	0.367	**0.885**					
(6) FB_OS_UP	0.472	0.428	0.306	0.302	0.554	**0.901**				
(7) MMQ	0.364	0.579	0.349	0.242	0.326	0.348	**0.855**			
(8) OSQ	0.401	0.695	0.307	0.342	0.463	0.449	0.552	**0.815**		
(9) PS	0.303	0.642	0.42	0.261	0.335	0.346	0.697	0.602	**0.919**	
(10) SC	0.244	0.484	0.441	0.241	0.329	0.292	0.399	0.365	0.513	**0.923**

Similarly, the results of *part 2* were summarized in Table 4 and Table 5. Overall, our measurement models were found to be appropriate for testing the hypothesized structural relationships.

Table 4. Convergent validity and internal consistency reliability for the 2nd-order constructs and remaining constructs

		Loadings	Weight	CR	AVE	α
HWP		1.000	1.000	0.895	0.870	1.000
SWQ		1.000	1.000	0.906	0.884	1.000
SERQ		1.000	1.000	0.877	0.842	1.000
CL	cl1	0.896	0.289	0.944	0.92	0.807
	cl2	0.909	0.273			
	cl3	0.897	0.268			
	cl4	0.892	0.283			
CS	cs1	0.922	0.374	0.941	0.906	0.842
	cs2	0.919	0.358			
	cs3	0.912	0.358			
PRA	pra1	0.841	0.301	0.908	0.866	0.713
	pra2	0.817	0.277			
	pra3	0.868	0.313			

(*continued*)

Table 4. (*continued*)

		Loadings	Weight	CR	AVE	α
	pra4	0.850	0.293			
PEA	pea1	0.802	0.333	0.903	0.856	0.699
	pea2	0.811	0.267			
	pea3	0.885	0.315			
	pea4	0.843	0.282			
Price	price1	0.881	0.250	0.933	0.905	0.778
	price2	0.877	0.292			
	price3	0.849	0.284			
	price4	0.920	0.307			
Design	design1	0.832	0.262	0.927	0.895	0.761
	design2	0.885	0.291			
	design3	0.892	0.299			
	design4	0.879	0.294			

Table 5. Discriminant validity for the 2nd-order constructs and remaining constructs

	(1)	(2)	(3)	(4)	(5)	(6)	(7)	(8)	(9)
(1) CL	**0.898**								
(2) Design	0.643	**0.872**							
(3) HWP	0.630	0.585	**1.000**						
(4) CS	0.825	0.643	0.660	**0.918**					
(5) PEA	0.781	0.644	0.654	0.803	**0.836**				
(6) PRA	0.715	0.643	0.651	0.767	0.773	**0.844**			
(7) Price	0.450	0.333	0.312	0.401	0.418	0.357	**0.882**		
(8) SERQ	0.481	0.371	0.484	0.470	0.503	0.536	0.489	**1.000**	
(9) SWQ	0.659	0.572	0.686	0.687	0.672	0.733	0.437	0.652	**1.000**

5.3 Test of Structural Model

Bootstrapping with 500 resamples was performed on both two parts of the measurement model to get estimates of standard errors for testing the statistical significance of path coefficients using t-statistics. Because our model contains the high-order constructs, we obtained latent variable scores for lower-order latent variables, and subsequently use them as manifest variables for the higher-order latent variables.

The results of our structural model are summarized in Table 6. We estimated three models. Model 1 is the full model, and Model 2 is a model with only the theoretical variables (excluding control variables) as predictors. Model 3 is a controls-only model that provides a benchmark for assessing the additional impacts of the theoretical variables.

In Model 1, all of the hypotheses were supported according to the statistical results. A comparison of Model 1 and 3 shows that the full model explains a substantive incremental variance of 22.4% (72.1%–49.7%). Out of the control variables, price, design, and age have significant impacts on customer loyalty. Moreover, the inclusion of the control variables only explains an additional 4% (72.1%–68.1%) of the variance (Model 1 and 2). These results suggest that our theoretical model is sufficiently substantive to explain a large proportion of the variance in CL.

Table 6. Results of structural model testing.

Hypothesis	Supported	Model 1	Model 2	Model 3
H1: CS → CL	Yes	0.648***	0.825***	
H2: PRA → CS	Yes	0.365***	0.365***	
H3: PEA → CS	Yes	0.521***	0.521***	
H4a: HWP → PRA	Yes	0.274***	0.274***	
H4b: HWP → PEA	Yes	0.36***	0.36***	
H5a: SWQ → PRA	Yes	0.491***	0.491***	
H5b: SWQ → PEA	Yes	0.366***	0.366***	
H6a: SERQ → PRA	Yes	0.084*	0.084*	
H6b: SERQ → PEA	Yes	0.09*	0.09*	
Price		0.121***		0.253***
Design		0.175***		0.532***
Gender		−0.036		−0.044
Age		0.041*		0.074**
Education		−0.019		−0.036
Experience		0.007		0.047
Annual Income		−0.015		−0.016
Primary Objectives		0.050		0.102***
R^2		72.1	68.1	49.7
Adjusted R^2		71.7	68.1	49.1

*: p<0.05, **: p<0.01, ***: p<0.001

5.4 Brand Comparison

To see the difference among the three smartphone brands, additional analyses were conducted by separating the data in terms of the brands. Table 7 shows the group comparison results.

We can see that for Samsung's customers, the impact of HWP on PRA is weaker than those for Xiaomi and Apple's customers. On the other hand, the effects of SWQ on PRA and PEA for Samsung's customers are higher than those for Xiaomi and Apple's customers. The results are the opposite of our initial expectation, given that Samsung is known as a hardware-oriented mobile phone manufacturer whereas Xiaomi is known to be relatively software-oriented. The reason for the unexpected results could be that, when we test a second order model, the first-order factors may have different values.

Accordingly, we tested another model by removing the second-order constructs (i.e., HWP, SWP, and SERQ) and directly linked the first order constructs with PRA and PEA. Internal reliability, convergent validity and discriminant validity were all checked in the same method with the full model.

Table 7. Brand comparison (the full model, Model 1)

Hypothesis	Xiaomi (N = 247)	Samsung (N = 218)	Apple (N = 245)
H1: CS → CL	0.662(11.224)***	0.614(7.781)***	0.553(7.342)***
H2: PRA → CS	0.289(5.222)***	0.280(3.308)**	0.526(6.585)**
H3: PEA → CS	0.584(11.353)***	0.625(7.783)***	0.329(4.491)***
H4a: HWP → PRA	0.264(3.323)**	0.201(3.074)**	0.291(4.686)***
H4b: HWP → PEA	0.363(5.213)***	0.331(3.694)***	0.252(3.767)***
H5a: SWQ → PRA	0.344(3.697)***	0.501(7.022)***	0.488(6.815)***
H5b: SWQ → PEA	0.143(1.930)	0.489(5.787)***	0.390(4.880)***
H6a: SERQ → PRA	0.258(4.044)***	0.177(2.901)**	0.046(0.651)
H6b: SERQ → PEA	0.337(5.564)***	0.001(0.003)	0.170(2.219)*
Price	0.106(1.891)	0.190(4.312)***	0.174(3.389)**
Design	0.1419(2.358)*	0.121(1.752)	0.211(2.940)**
Gender	−0.070(1.876)	0.014(0.425)	−0.044(1.016)
Age	0.012(0.297)	0.090(2.673)**	0.028(0.934)
Edu	−0.084(2.187)*	−0.019(0.548)	0.012(0.297)
Income	0.007(0.181)	−0.020(0.499)	−0.009(0.196)
Experience	−0.010(0.270)	0.006(0.189)	0.019(0.457)
Primary objectives	0.073(1.816)	0.052(1.489)	0.054(0.984)

*: $p<0.05$, **: $p<0.01$, ***: $p<0.001$

As seen in Table 8, Samsung has the least impact of BP on PRA and PEA, and Apple has the lowest level of effect of SC on PRA and PEA. Apple has the strongest effect of PS on PRA, and Xiaomi has the largest impact of PS on PEA. Xiaomi is relatively weak in the effect of MMQ on PEA, which is known as the conventional advantage of Samsung and Apple. In addition, the impacts of FB_OS_UP on PRA and PEA are found to be greatest for Xiaomi, compared to the other two brands. The three brands have similar effects of AFT_SERQ on PRA, while Xiaomi has the highest impact of AFT_SERQ on PEA. Lastly, Samsung shows weak in COMS_SERQ while Apple is weak in COMO_SERQ.

Table 8. Brand comparison without 2nd-order constructs

	Xiaomi (N = 247)	Samsung (N = 218)	Apple (N = 245)
HWP	Path coefficient (T value)		
BP → PRA	0.195(3.159)**	0.037(0.665)	0.058(0.985)
BP → PEA	0.205(3.236)**	0.100(1.566)	0.238(3.583)***
SC → PRA	0.109(1.484)	0.194(2.42)*	0.018(0.342)
SC → PEA	0.055(0.729)	0.155(2.154)*	0.019(0.301)
PS → PRA	0.278(3.203)**	0.335(3.851)***	0.454(5.76)***
PS → PEA	0.431(5.236)***	0.347(4.098)***	0.31(4.169)***
MMQ → PRA	0.217(3.105)**	0.212(2.263)*	0.246(3.371)**
MMQ → PEA	0.096(1.375)	0.228(2.842)**	0.194(2.811)**
SWQ	Path coefficient (T value)		
FB_OS_UP → PRA	0.117(2.133)*	0.104(2.298)*	0.098(1.775)
FB_OS_UP → PEA	0.269(4.447)***	0.226(3.846)***	0.239(3.258)**
APPQ → PRA	0.423(6.127)***	0.467(7.354)***	0.398(5.333)***
APPQ → PEA	0.275(3.004)**	0.39(5.121)***	0.255(2.955)**
OSQ → PRA	0.285(4.324)***	0.335(5.298)***	0.338(4.759)***
OSQ → PEA	0.235(2.718)**	0.272(3.594)***	0.296(3.739)***
SERQ	Path coefficient (T value)		
AFT_SERQ → PRA	0.198(2.895)**	0.189(3.256)**	0.190(2.483)*
AFT_SERQ → PEA	0.298(5.092)***	0.185(2.842)**	0.203(2.66)**
COMS_SERQ → PRA	0.319(4.542)***	0.238(3.629)***	0.271(3.319)**
COMS_SERQ → PEA	0.287(3.993)***	0.202(3.101)**	0.300(3.929)***
COMO_SERQ → PEA	0.220(3.146)**	0.276(3.094)**	0.185(3.164)**
COMO_SERQ → PRA	0.319(4.382)***	0.399(5.562)***	0.168(3.065)**

*: $p<0.05$, **: $p<0.01$, ***: $p<0.001$

5.5 Additional Analysis: Switching Brands

As shown in Table 9, about 40% of current Xiaomi and Apple's users switched from Samsung. This raises the question of why they decided to switch. To address this question, we conducted an additional test on the users who switched from Samsung and switched to Samsung. Table 10 shows which of the first factors of our model has significant impact on PRA and PEA, both of which may lead to brand switching.

Comparing the previously Samsung users (N = 201 = 99 + 102) and currently Samsung users (N = 218), we can see that Samsung has a higher performance on SC, MMQ, FB_OS_UP, APPQ, COMS_SERQ, and COMO_SERQ. It also has a similar level of OSQ with Xiaomi and Apple. However, Samsung is found to be relatively weak in the effect of AFT_SERQ. In general, it appears that Samsung is quite competitive

Table 9. Switching brands

Xiaomi (N = 247)			Samsung (N = 218)			Apple (N = 245)		
Previous brand	Frequency (%)		Previous brand	Frequency (%)		Previous brand	Frequency (%)	
Samsung	102	41.3	Xiaomi	40	18.3	Xiaomi	34	13.9
Apple	28	11.3	Apple	55	25.2	Samsung	99	40.4
Huawei	39	15.8	Huawei	39	17.9	Huawei	27	11.0
Lenovo	32	13.0	Lenovo	35	16.1	Lenovo	27	11.0
Others	46	18.6	Others	49	22.5	Others	58	23.7

compared with Xiaomi and Apple. However, this does not clearly explain why more than 40% of current Xiaomi and Apple's users has switched from Samsung.

For this, we separated the data into three groups: users switched from Samsung to Xiaomi (N = 102), Samsung to Apple (N = 99), and Xiaomi/Apple to Samsung (N = 95 = 40 + 55). As Table 10 shows, when we see the first group (i.e., Samsung to Xiaomi (N = 102)), it can be inferred that Xiaomi outperforms Samsung in FB_OS_UP and AFT_SERQ in terms of PRA and PEA. In contrast, Samsung and Apple have a relatively advantage on PS over Xiaomi (i.e., comparison of the first three groups).

Table 10. The effects of the 1st-order factors on PRA and PEA under condition of switching

	Samsung to Xiaomi (N = 102)	Samsung to Apple (N = 99)	Xiaomi/Apple to Samsung (N = 95)	Previously Samsung (N = 201)	Currently Samsung (N = 218)
HWP	path coefficient (T value)				
BP → PRA	0.197(1.616)	0.036(0.537)	−0.028(0.337)	0.091(1.441)	0.037(0.653)
BP → PEA	0.092(0.791)	0.223(2.294)*	−0.024(0.272)	0.14(1.893)	0.1(1.578)
SC → PRA	0.225(2.126)*	0.014(0.204)	0.179(1.383)	0.073(1.343)	0.194(2.341)*
SC → PEA	0.119(1.188)	0.031(0.32)	0.143(1.118)	0.03(0.494)	0.155(2.004)*
PS → PRA	0.299(2.144)*	0.649(7.418)***	0.442(3.903)***	0.496(6.53)***	0.335(3.846)***
PS → PEA	0.445(3.826)***	0.312(2.439)*	0.459(4.221)***	0.447(5.731)***	0.347(3.943)***
MMQ → PRA	0.125(1.196)	0.16(1.856)	0.222(1.707)	0.197(2.74)**	0.212(2.128)*
MMQ → PEA	0.153(1.27)	0.194(2.02)*	0.235(2.12)*	0.166(2.069)*	0.228(2.694)**

(continued)

Table 10. (*continued*)

	Samsung to Xiaomi (N = 102)	Samsung to Apple (N = 99)	Xiaomi/Apple to Samsung (N = 95)	Previously Samsung (N = 201)	Currently Samsung (N = 218)
SWQ	path coefficient (T value)				
FB_OS_UP → PRA	0.181(2.259)*	0.073(0.803)	0.115(1.699)	0.103(1.794)	0.104(2.147)*
FB_OS_UP → PEA	0.294(3.031)**	0.175(1.728)	0.076(0.998)	0.162(2.392)*	0.226(3.741)***
APPQ → PRA	0.398(4.03)***	0.425(4.695)***	0.332(2.881)**	0.411(5.986)***	0.467(7.175)***
APPQ → PEA	0.299(3.174)**	0.461(4.161)***	0.554(6.095)***	0.372(4.873)***	0.39(4.862)***
OSQ → PRA	0.317(3.461)**	0.32(3.466)**	0.276(2.916)**	0.35(5.309)***	0.335(5.469)***
OSQ → PEA	0.253(2.765)**	0.099(0.785)	0.458(3.982)***	0.232(2.776)**	0.272(3.435)**
SERQ	path coefficient (T value)				
AFT_SERQ → PRA	0.432(4.569)***	0.161(1.406)	0.219(2.633)**	0.317(3.775)***	0.189(3.289)**
AFT_SERQ → PEA	0.51(5.232)***	0.146(1.281)	0.232(2.479)*	0.361(4.271)***	0.185(2.764)**
COMS_SERQ → PRA	0.232(2.186)*	0.259(1.844)	0.406(4.106)***	0.178(1.481)	0.238(3.721)***
COMS_SERQ → PEA	0.238(2.138)*	0.22(1.654)	0.332(3.179)**	0.146(1.319)	0.202(2.848)**
COMO_SERQ → PRA	0.186(1.763)	0.122(1.155)	0.33(3.265)**	0.075(0.903)	0.399(5.918)***
COMO_SERQ → PEA	0.021(0.178)	0.179(1.674)	0.302(2.634)**	0.025(0.303)	0.276(3.377)**

6 Contributions and Limitations

Our study extends the extant knowledge about CS&L to the context of smartphone usage. Theoretically, our results provide evidence that the cognitive-affective system theory of personality (CAPS) is useful for explaining and understanding smartphone users' CS&L. To be specific, our model suggests that when using CAPS to understand smartphone usage behavior, users' cognitive and affective beliefs can be specifically measured with the perceived relative advantage variable and the perceived emotional attachment variable, respectively.

Further, our model theoretically links the recent competing factors for smartphone companies, i.e., hardware, software, and service, with the overarching theory of our study, i.e., CAPS, as the antecedents of the cognitive and affective beliefs. Our results explain how smartphone companies' strategies related to the competing points can generate users' cognitive and affective reactions, which in turn affect users' CS&L. This contributes to the literature on CS&L by providing knowledge that cognitive and affective beliefs of smartphone users can be formed or led by smartphone companies' strategic differences on hardware, software, and service.

Our study also provides practitioners with an opportunity to evaluate and further improve their current strategies for CS&L in terms of hardware performance, software quality, and service quality. For example, it is known that Xiaomi takes much care of customer satisfaction by improving the quality of after-sale service. The company promises customers that they try to handle and solve their product-related customer troubles within half an hour. Given that the company outperforms the other competitors in terms of the quality of after-sale service in our study, this study provides empirical evidence that the company's strategy works well at the moment at least.

Along similar lines, our study also lends an industry-wide screenshot about how the leading smartphone companies manage the three competing components of smartphones to achieve higher levels of user satisfaction and loyalty. Our comparisons of the three brands allow practitioners to understand that CS&L of smartphone users can be obtained in many different manners, and that diverse strategies from different combinations of the three competing components may be the key for their sustainable comepetitive edge.

Although our proposed model was supported by the data, our study also has some limitations. For example, a large portion of our data came from young generation in China, missing the variance of customers at other ages. Those missed customers may also be an important user category for smartphone businesses. Future research may extend our study through further data collection and may compare Chinese smartphone markets with other countries such as the U.S. market.

References

1. Aydin, S., Özer, G.: The analysis of antecedents of customer loyalty in the Turkish mobile telecommunication market. Eur. J. Mark. **39**, 910–925 (2005)
2. Bae, J.K., Jeong, H.M.: An empirical study on the determinants factors by including functional attributes of smart phone adoption. e-Bus. Stud. **9**, 337–361 (2008)
3. Ball, A.D., Tasaki, L.H.: The role and measurement of attachment in consumer behavior. J. Consum. Psychol. **1**, 155–172 (1992)
4. Bagozzi, R.P., Dholakia, U.M.: Antecedents and purchase consequences of customer participation in small group brand communities. Int. J. Res. Mark. **23**, 45–61 (2006)
5. Balabanis, G., Reynolds, N., Simintiras, A.: Bases of e-store loyalty: perceived switching barriers and satisfaction. J. Bus. Res. **59**, 214–224 (2006)
6. Barclay, D., Thompson, R., Higgins, C.: The partial least squares (PLS) approach to causal modeling: personal computer adoption and use as an illustration. Technol. Stud. **2**, 285–309 (1995)
7. Bellis, M.: History of the electric battery. About.com (2008)
8. Bowlby, J.: Loss: Sadness and Depression. Basic Books, New York (1980)
9. Caro, L.M., García, J.A.M.: Cognitive–affective model of consumer satisfaction: an exploratory study within the framework of a sporting event. J. Bus. Res. **60**, 108–114 (2007)
10. Chang, Y.F., Chen, C.S., Zhou, H.: Smart phone for mobile commerce. Comput. Stand. Interfaces **31**, 740–747 (2009)
11. Chang, Y.H., Kim, M.K., Kim, Y.E., Park, M.C.: Determinants of user satisfaction and continuance intention of smartphones: focus on interactivity perspective. In: Proceedings of the 8th International Telecommunications Society (ITS) Asia-Pacific Regional Conference, pp. 258–271 (2011)
12. Chin, W.W.: The partial lease squares approach for structural equation modeling. In: Modern Methods for Business Research (1998)

13. Choudhury, V., Karahanna, E.: The relative advantage of electronic channels: a multidimensional view. MIS Q. **32**, 179–200 (2008)
14. Churchill Jr., G.A., Surprenant, C.: An investigation into the determinants of customer satisfaction. J. Mark. Res. **19**(4), 491–504 (1982)
15. Cyr, D., Head, M., Ivanov, A.: Design aesthetics leading to m-loyalty in mobile commerce. Inf. Manag. **43**, 950–963 (2006)
16. Ding, X.Q.: Understanding Customer Loyalty in Smartphone Market: Comparison of Xiaomi, Samsung, and Apple Phone Users, Master's Thesis, School of Business, Yonsei University (2015)
17. Fornell, C., Bookstein, F.L.: Two structural equation models: LISREL and PLS applied to consumer exit-voice theory. J. Mark. Res. **19**(4), 440–452 (1982)
18. Grisaffe, D.B., Nguyen, H.P.: Antecedents of emotional attachment to brands. J. Bus. Res. **64**, 1052–1059 (2011)
19. Guo, L., Xiao, J.J., Tang, C.: Understanding the psychological process underlying customer satisfaction and retention in a relational service. J. Bus. Res. **62**, 1152–1159 (2009)
20. Haverila, M.: Mobile phone feature preferences, customer satisfaction and repurchase intent among male users. Australas. Mark. J. **19**, 238–246 (2011)
21. Hong, W., Thong, J.Y.L., Chasalow, L.C., Dhillon, G.: User acceptance of agile information systems: a model and empirical test. J. Manag. Inf. Syst. **28**, 235–272 (2011)
22. Johnston, R.: The determinants of service quality: satisfiers and dissatisfiers. Int. J. Serv. Ind. Manag. **6**, 53–71 (1995)
23. Kang, Y.M., Cho, C.W., Lee, S.J.: Analysis of factors affecting the adoption of smartphones. In: Proceedings of the IEEE International Technology Management Conference, pp. 919–925 (2011)
24. Kim, H.J., Agrawal, N., Ungureanu, C.: Revisiting storage for smartphones. ACM Trans. Storage **8**, 1–25 (2012)
25. Kim, H.J., Kim, I.C., Lee, H.G.: Third-party mobile app developers' continued participation in platform-centric ecosystems: an empirical investigation of two different mechanisms. Int. J. Inf. Manag. **36**, 44–59 (2016)
26. Kim, H.S., Yoon, C.H.: Determinants of subscriber churn and customer loyalty in the korean mobile telephony market. Telecommun. Policy **28**, 751–765 (2004)
27. Kim, Y.L., Briley, D.A., Ocepek, M.G.: Differential innovation of smartphone and application use by sociodemographics and personality. Comput. Hum. Behav. **44**, 141–147 (2015)
28. Koch, S., Kerschbaum, M.: Joining a smartphone ecosystem: application developers' motivations and decision criteria. Inf. Softw. Technol. **56**, 1423–1435 (2014)
29. Lai, F., Griffin, M., Babin, B.J.: How quality, value, image, and satisfaction create loyalty at a Chinese telecom. J. Bus. Res. **62**, 980–986 (2009)
30. Lee, Y., Kwon, O.: Intimacy, familiarity and continuance intention: an extended expectation-confirmation model in web-based services. Electron. Commer. Res. Appl. **10**, 342–357 (2011)
31. Lee, J.W.: Critical factors promoting customer loyalty to smartphone and mobile communications service providers. Acad. Mark. Stud. J. **15**(1), 59–69 (2011)
32. Leong, L.Y., Hew, T.S., Lee, V.H., Ooi, K.B.: An SEM–artificial-neural-network analysis of the relationships between SERVPERF, customer satisfaction and loyalty among low-cost and full-service airline. Expert Syst. Appl. **42**(19), 6620–6634 (2015)
33. Liang, D.P., Ma, Z.Z., Qi, L.Y.: Service quality and customer switching behavior in China's mobile phone service sector. J. Bus. Res. **66**, 1161–1167 (2013)
34. Lin, W.B.: The exploration of customer satisfaction model from a comprehensive perspective. Expert Syst. Appl. **33**, 110–121 (2007)
35. Mano, H., Oliver, R.L.: Assessing the dimensionality and structure of the consumption experience: evaluation, feeling, and satisfaction. J. Consum. Res. **20**(3), 451–466 (1993)

36. Mansfield, I.: Smartphone battery life has become a significant drain on customer satisfaction and loyalty (2012)
37. Mischel, W., Shoda, Y.: A cognitive-affective system theory of personality: reconceptualizing situations, dispositions, dynamics, and invariance in personality structure. Psychol. Rev. **102**, 246–268 (1995)
38. Mischel, W.: Toward a cognitive social learning reconceptualization of personality. Psychol. Rev. **80**, 252–283 (1973)
39. MIT Electric Vehicle Team: A guide to understanding battery specifications (2008)
40. Moore, G.C., Benbasat, I.: Development of an instrument to measure the perceptions of adopting an information technology innovation. Inf. Syst. Res. **2**, 192–222 (1991)
41. Nelson, R.R., Todd, P.A., Wixom, B.H.: Antecedents of information and system quality: an empirical examination within the context of data warehousing. J. Manag. Inf. Syst. **21**, 199–235 (2005)
42. Oliver, R.L.: A cognitive model of the antecedents and consequences of satisfaction decisions. J. Mark. Res. **17**(4), 460–469 (1980)
43. Oliver, R.L.: Cognitive, affective, and attribute bases of the satisfaction response. J. Consum. Res. **20**(3), 418–430 (1993)
44. Parasuraman, A., Zeithaml, V.A., Berry, L.L.: A conceptual model of service quality and its implications for future research. J. Mark. **49**, 41–50 (1985)
45. Pi, S.M., Liao, H.L., Chen, H.M.: Factors that affect consumers' trust and continuous adoption of online financial services. Int. J. Bus. Manag. **7**, 108–119 (2012)
46. Ray, S., Kim, S.S., Morris, J.G.: Research note-online users' switching costs: their nature and formation. Inf. Syst. Res. **23**, 197–213 (2012)
47. Rhee, H.S., Riggins, F.J., Kim, C.: The impact of product type and perceived characteristics of the web on multifaceted online shopping behavior. J. Organ. Comput. Electron. Commer. **19**, 1–29 (2009)
48. Rogers, E.: Diffusion of Innovations, 4th edn. Free Press, New York (1995)
49. Sanchez-Franco, M.J., Roldán, J.L.: Expressive aesthetics to ease perceived community support: exploring personal innovativeness and routinised behaviour as moderators in Tuenti. Comput. Hum. Behav. **26**, 1445–1457 (2010)
50. Sampson, S.E., Showalter, M.J.: The performance-importance response function: observations and implications. Serv. Ind. J. **19**, 1–25 (1999)
51. Shin, D.H.: Effect of the customer experience on satisfaction with smartphones: assessing smart satisfaction index with partial least squares. Telecommun. Policy **39**(8), 627–641 (2014)
52. Sperry, J.: Don't just collect customer feedback: listen, engage and act (2014). http://www.business2community.com/customer-experience/dont-just-collect-customer-feedback-listen-engage-act-0776235
53. Stuedi, P., Mohomed, I., Terry, D.: WhereStore: location-based data storage for mobile devices interacting with the cloud, mobile cloud computing & services. In: Proceedings of the 1st ACM Workshop, pp. 1–8 (2010)
54. Swan, J.E., Combs, L.J.: Product performance and consumer satisfaction: a new concept. J. Mark. **40**(2), 25–33 (1976)
55. Thomson, M., MacInnis, D.J., Park, C.W.: The ties that bind: measuring the strength of consumers' emotional attachments to brands. J. Consum. Psychol. **15**, 77–91 (2005)
56. Wang, Y.G., Ma, S.S., Li, D.H.: Customer participation in virtual brand communities: the self-construal perspective. Inf. Manag. **52**(5), 577–587 (2015)
57. Yini, H.: China's mobile users hit 1.3 billion in 2015. Chinadaily.com (2016)
58. Zeithaml, V.A., Berry, L.L., Parasuraman, A.: The behavioral consequences of service quality. J. Mark. **60**, 31–46 (1996)

Check for updates

Vertical Data Mining from Relational Data and Its Application to COVID-19 Data

Pranjal Gupta[1,2,3], Calvin S. H. Hoi[2], Carson K. Leung[2(✉)], Ye Yuan[2], Xiaoke Zhang[2,4,5], and Zhida Zhang[2,6,7]

[1] Birla Institute of Technology and Science (BITS) - Pilani, Jhunjhunu, RJ, India
[2] University of Manitoba, Winnipeg, MB, Canada
kleung@cs.umanitoba.ca
[3] University of Waterloo, Waterloo, ON, Canada
[4] Nanjing University, Nanjing, JS, China
[5] The University of British Columbia (UBC), Vancouver, BC, Canada
[6] Tongji University, Shanghai, China
[7] Columbia University, New York, NY, USA

Abstract. Nowadays, big data are everywhere. These big data can be of different degrees of veracity (e.g., precise, imprecise and uncertain data). Many of them are open data and are stored in relational databases. Embedded in these big data are valuable information and knowledge, which can be discovered by data mining. Frequent pattern mining is a popular data mining task within the realm of data science. In this paper, we present data science techniques for vertical data mining—in particular, vertical mining of frequently occurring patterns—from these relational data. For illustration, we discuss applications of the vertical data mining for the discovery of knowledge and useful information from real-life epidemiological data about coronavirus disease 2019 (COVID-19) and economic data.

Keywords: Data mining · Data science · Frequent patterns · Relational data · Big data applications · COVID-19

1 Introduction and Related Works

As we are living in the era of *big data* [1–5], big data are everywhere. Specifically, advancements in information and communication technology (ICT) have led to easy generation and collection of huge amounts of valuable data at a rapid rate. Due to the inherited nature of the instructions for generating and measuring data, these big data can be of different degrees of veracities. For instance, some are precise, while others are imprecise and uncertain [6–8].

Embedded in these big data are implicit, previously unknown and potentially useful information and valuable knowledge. This calls for *data science* [9–13]—which applies database techniques, scientific and engineering methods, mathematical and statistical models, machine learning (e.g., deep learning), and/or data mining—to discover the valuable knowledge. As a common data mining task, *frequent pattern mining* [14–16]

W. Lee et al. (Eds.): *Big Data Analyses, Services, and Smart Data*, AISC 899, pp. 106–116, 2021.
https://doi.org/10.1007/978-981-15-8731-3_8

aims to find sets (or collections) frequently co-occurring items (e.g., bundles of popular merchandise items in shopper market baskets, collections of frequently collocated conferences/events). It has been applied to numerous real-life applications and services (e.g., social network analysis [17–20]). Since its introduction [14], many frequent pattern mining algorithms [21–28] have been proposed to first find frequent patterns and then use the mined frequent patterns to form interesting association rules. Most of the algorithms mine frequent patterns horizontally in the sense that they mine transaction-centric data. For example, from transactional data in Fig. 1, the algorithms find a frequent pattern {*a*, *b*, *c*}.

Transaction ID	Transaction content
t1	*a, b, c*
t2	*a, b, c, d, e*
t3	*a, b, c*

Fig. 1. A sample of transactional data

In recent years, due to the popularity of the concept of *open data* [29, 30], there has been a trend for more collections of big data to be made openly available by scientists, researchers, and non-profit organizations. Moreover, many governments also have made their collected big data freely available to their citizens. Some examples include:

- Yellow (i.e., medallion) taxi trip records[1], green (i.e., street hail livery) taxi trip records, and for-hire vehicle (FHV) trip records collected and shared by New York City (NYC) Taxi & Limousine Commission (TLC) as a part of the NYC Open Data[2];
- Transit (i.e., bus) on-time performance data[3] collected and shared by Winnipeg Transit as a part of the City of Winnipeg Open Portal[4]; as well as
- Information on cases of coronavirus disease (COVID-19) collected from epidemiological reports [31–33] from provincial/territorial partners via Public Health Agency of Canada and shared by Statistics Canada as a part of Canadian Open Government Portal[5].

It was observed that, while some of these open data are stored as semi-structured data in JavaScript Object Notation (JSON), many of these open data are stored as structured, relational data and can be downloaded in comma-separated values (CSV) files. In other words, the transactional data shown in Fig. 1 can be represented in a relational form as shown in Fig. 2.

As a concrete example, a dataset on COVID-19 cases is stored in relational form such that a case may be captured by multiple rows and each row captures the value for a particular attribute of a case. See Fig. 3.

1 https://www1.nyc.gov/site/tlc/about/tlc-trip-record-data.page.
2 https://opendata.cityofnewyork.us/.
3 https://winnipegtransit.com/en/open-data/on-time-performance/.
4 https://data.winnipeg.ca/.
5 https://open.canada.ca/.

Transaction ID	Item
t1	*a*
t1	*b*
t1	*c*
t2	*a*
t2	*b*
t2	*c*
t2	*d*
t2	*e*
t3	*a*
t3	*b*
t3	*c*

Fig. 2. A sample of relational data

Case ID	Attribute	Value
1	region	Quebec
1	gender	male
1	age group	30-39
1	...	...
1	transmission	domestic acquisition
2	...	...

Fig. 3. A sample of COVID-19 case data

For these relational data, it seems logical to mine frequent patterns *vertically* in the sense that they mine item-centric data or (*attribute, value*)-centric data. Vertical frequent pattern mining can be considered as an alternative to horizontal mining.

In the remainder of this paper, we describe vertical frequent pattern mining from precise data in Sect. 2. Then, we describe vertical frequent pattern mining from uncertain data in Sects. 3 and 4. In Sect. 5, we illustrate the key concepts by applying vertical frequent pattern mining to two real-life datasets, including epidemiological data on COVID-19 cases. Finally, we draw our conclusions in Sect. 6.

2 Vertical Frequent Pattern Mining from Precise Data

2.1 The Eclat Algorithm

To mine frequent patterns vertically, the Eclat algorithm [23] applies a breadth-first (i.e., level-wise) bottom-up approach. It first scans the dataset to find a set of transaction ID (tID) for each item. The resulting set is a tID set for each item. The size of the tID set for item x shows the frequency (i.e., actual support) of the pattern $\{x\}$ of size 1 (i.e., 1-itemset). If the frequency of $\{x\}$ is not less than the user-specified support threshold *minsup*, then $\{x\}$ is considered frequent.

Set cardinality $k = 1$. By applying the set intersection operation on the tID sets of two frequent k-itemsets X and Y (where X and Y shared common (k – 1) "prefix" items), the algorithm obtains the intersected tID set—which is the tID set of (k + 1)-itemset

$(X \cup Y)$:

$$\text{tID set of } (X \cup Y) = (\text{tID set of } X) \cap (\text{tID set of } Y) \quad (1)$$

Again, the size of the tID set shows the frequency:

$$\sup(X \cup Y) = |\text{tID set of } (X \cup Y)| \quad (2)$$

If $\sup(X \cup Y) \geq minsup$, then $(X \cup Y)$ is frequent. This set intersection and set size operations, as listed in Eqs. (1) and (2), are applied repeatedly for $k \geq 2$ for find all frequent patterns of higher cardinality k.

2.2 The dEclat Algorithm

Depending on the density of the datasets, the size of tID set of some k-itemsets (especially for low cardinality k) can be large for dense datasets. So, for these dense datasets, the dEclat algorithm [24] scans the dataset to find the complement of the tID set for each item. The resulting set is a diffset for each item. It captures the IDs of transactions in which the item is *absent* (cf. tID set captures the IDs of transactions in which the item is *present*). Note that, for dense dataset, such a diffset is expected to be small when compared to the large size of the corresponding tID set. The difference between the total number of transactions in the dataset and the size of the diffset of an item x gives the frequency (i.e., actual support) of 1-itemset $\{x\}$. If $\sup(\{x\}) \geq minsup$, then $\{x\}$ is frequent.

Set cardinality $k = 1$. By applying the set difference operation on the tID sets of two frequent k-itemsets Y and X (where X and Y shared common $(k - 1)$ "prefix" items), the algorithm obtains the diffset of $(k + 1)$-itemset $(X \cup Y)$:

$$\text{diffset of } (X \cup Y) = (\text{diffset of } Y) - (\text{diffset of } X) \quad (3)$$

The frequency of $(X \cup Y)$ can be computed by subtracting the size of the diffset of $(X \cup Y)$ from the support of X:

$$\sup(X \cup Y) = \sup(X) - |\text{diffset of } (X \cup Y)| \quad (4)$$

If $\sup(X \cup Y) \geq minsup$, then $(X \cup Y)$ is frequent. This set difference and subtraction operations, as listed in Eqs. (3) and (4), are applied repeatedly for $k \geq 2$ for find all frequent patterns of higher cardinality k.

2.3 The VIPER Algorithm

Besides the tID set and diffset representations, the datasets can also be captured by bit vector (or bitmap, binary matrix) with bits "0" and "1". For these datasets, the VIPER algorithm [25] first scans the dataset to find a bit vector for each item. The number of "1" in the resulting bit vector for item x shows the frequency of the 1-itemset $\{x\}$. If $\sup(\{x\}) \geq minsup$, then $\{x\}$ is frequent.

Set cardinality $k = 1$. By applying the dot product on the bit vectors of two frequent k-itemsets X and Y (where X and Y shared common ($k - 1$) "prefix" items), the algorithm obtains the bit vector of ($k + 1$)-itemset (X ∪ Y):

$$\text{bit vectorof } (\text{X} \cup \text{Y}) = (\text{bit vector of X}) \bullet (\text{bit vector of Y}) \tag{5}$$

Again, the number of "1" in the resulting bit vector for (X ∪ Y) gives sup(X ∪ Y). If sup(X ∪ Y) ≥ *minsup*, then (X ∪ Y) is frequent. This dot product and counting of "1" operations, as listed in Eq. (5), are applied repeatedly for $k \geq 2$ for find all frequent patterns of higher cardinality k. Note that the size of bit vector is dependent on the number of transactions (and items) but not the density of the dataset.

2.4 A Hybrid Algorithm

To speed up the vertical mining process, a hybrid algorithm [26] takes advantages of these three algorithms and switch from one to another depending on the memory requirement and density of the datasets. Specifically, it uses with Eclat when data are sparse data, dEclat when data are dense, and VIPER when tID sets or diffsets occupy more than 1/8 of the total bits for N transactions and M items (i.e., MN ÷ 8).

3 Vertical Frequent Pattern Mining from Uncertain Data

3.1 The UV-Eclat Algorithm

To mine frequent patterns from these uncertain data, the UV-Eclat algorithm [27] first scans the dataset to find the tID set for each item. As each tID is associated with an existential probability P(x, t) to indicate its likelihood of x appearing in transaction t in the uncertain data, the sum of these existential probability values gives the expected support of 1-itemset $\{x\}$. If it is not less than the user-specified support threshold *minsup* (i.e., expSup($\{x\}$) ≥ *minsup*), then $\{x\}$ is frequent.

Set cardinality $k = 1$. By applying the set intersection operation on the tID sets of a frequent k-itemset X and a frequent 1-itemset $\{y\}$, the algorithm obtains the intersected tID set—which is the tID set of ($k + 1$)-itemset (X ∪ $\{y\}$):

$$\text{tID set of } (\text{X} \cup \{y\}) = (\text{tID set of X}) \cap (\text{tID set of}\{y\}) \tag{6}$$

The product of the expected support expSup(X, t) of X in t and the existential probability P($\{y\}$, t) gives the expected support expSup(X ∪ $\{y\}$, t) of (X ∪ $\{y\}$) in t. Summing these products over all transactions gives the expected support expSup(X ∪ $\{y\}$) in the dataset:

$$\text{expSup}(\text{X} \cup \{y\}) = \sum_t (\text{expSup}(\text{X}, t) \times \text{P}(\{y\}, t)) \tag{7}$$

If expSup(X ∪ $\{y\}$) ≥ *minsup*, then (X ∪ $\{y\}$) is frequent. This set intersection operation for tIDs and sum of product of expSup(X, t) and P($\{y\}$, t), as listed in Eqs. (6) and (7), are applied repeatedly for $k \geq 2$ for find all frequent patterns of higher cardinality k.

3.2 The U-VIPER Algorithm

Similarly, the U-VIPER algorithm [28] first scans the dataset to find the bit vector for each item. The sum of these existential probability values associated with the "1" bit in the bit vector gives the expected support of 1-itemset $\{x\}$. If expSup($\{x\}$) $\geq$ *minsup*, then $\{x\}$ is frequent.

Set cardinality $k = 1$. By applying the dot product on the bit vectors of a frequent k-itemset X and a frequent 1-itemset $\{y\}$, the algorithm obtains the bit vector of (k + 1)-itemset (X ∪ $\{y\}$):

$$\text{bit vector of } (X \cup \{y\}) = (\text{bit vector of } X) \bullet (\text{bit vector of } \{y\}) \tag{8}$$

The product of the expected support expSup(X, t) of X in t and the existential probability P($\{y\}, t$) gives the expected support expSup(X ∪ $\{y\}, t$) of (X ∪ $\{y\}$) in t. Summing these products over all transactions gives the expected support expSup(X ∪ $\{y\}$) in the dataset. If expSup(X ∪ $\{y\}$) $\geq$ *minsup*, then (X ∪ $\{y\}$) is frequent. This dot product of bit vectors for tIDs and sum of product of expSup(X, t) and P($\{y\}, t$), as listed in Eq. (8), are applied repeatedly for $k \geq 2$ for find all frequent patterns of higher cardinality k.

4 Improvements to Vertical Frequent Pattern Mining from Uncertain Data

4.1 An Improved UV-Eclat Algorithm

Let X = (W ∪ $\{x\}$) and Y = (W ∪ $\{y\}$) be two k-itemsets sharing a common subset W. Observed that the UV-Eclat algorithm intersects frequent k-itemset X with any frequent 1-itemset $\{y\}$ to form (k + 1)-itemset Z = (X ∪ $\{y\}$). Although {y} is frequent, there is no guarantee that Y would be frequent. For any infrequent Y = W ∪ $\{y\}$, its superset Z = (X ∪ $\{y\}$) = (W ∪ $\{x, y\}$) would guarantee to be infrequent.

To address this issue, we intersect X with a frequent Y (rather than intersecting X with $\{y\}$) where X and Y share a common "prefix". Then, we compute its expected support expSup(X ∪ Y) by

$$\text{expSup}(X \cup Y) = \sum_t (\text{expSup}(X, t) \times P(\{y\}, t)) \tag{9}$$

where Y = (W ∪ $\{y\}$) such that Y shared the common "prefix" W with X = (W ∪ $\{x\}$). Hence, during the vertical mining process of a (k + 1)-itemset, we keep tID sets (and their existential probability values) for frequent k-itemsets and frequent 1-itemsets. This memory requirement is identical to that for the UV-Eclat algorithm, but we now reduce the number of set intersections (and only intersect those frequent X and Y).

4.2 An Improved U-VIPER Algorithm

We observed a similar issue on the U-VIPER algorithm. To address it, we apply dot product on X with a frequent Y (rather than on X with $\{y\}$) where X and Y share a common "prefix". Then, we compute its expected support expSup(X ∪ Y) by Eq. (9). Again, this memory requirement is identical to that for the U-VIPER algorithm, but we now reduce the number of dot products (and only apply to those frequent X and Y).

5 Case Studies: Vertical Mining from Relational Data

5.1 Mining Epidemiological Data on COVID-19 Cases

To illustrate and demonstrate the vertical frequent pattern mining algorithms described in this paper, we conducted a case study on the cases of coronavirus disease (COVID-19) collected from epidemiological reports from provincial/territorial partners via Public Health Agency of Canada and shared by Statistics Canada as a part of Canadian Open Government Portal. The information on these COVID-19 cases were stored in a CSV file in the format as shown in Fig. 3. This include:

- Administrative information (e.g., region),
- Case details (e.g., gender, age group, occupation),
- Symptoms (e.g., cough, fever, chills, short throat),
- Clinical course and outcomes (e.g., hospital status, recovery week), and
- Exposures (e.g., transmission).

By applying vertical frequent pattern mining to these data updated July 23, 2020 (for Canadian COVID-19 cases up July 22, 2020), we found the following frequent 1-itemsets:

- {transmission = 'domestic acquisition'} with a frequency of 89,498—which reveals that, among 101,121 confirmed cases in Canada (up to July 22, 2020), 89,498 cases (i.e., 88.5% of all confirmed cases) were reported to be community exposures. These include cases who reported having close contact with either (a) a confirmed or probable COVID-19 case in the 14 days prior to symptom onset or (b) a symptomatic person who had traveled to an affected area in the 14 days prior to their illness onset.
- {death = 'no'} with a frequency of 60,979—which reveals that 60,979 cases (i.e., 60.3% of all confirmed cases) were reported to survive and not deceased. The remaining 39.7% were either not stated with any death status (31.4%) or deceased (8.3%).
- {hospital status = 'not hospitalized'} with a frequency of 57,640—which reveals that 57,640 cases (i.e., 57.0% of all confirmed cases) were reported not to be hospitalized. The remaining 43.0% were not stated with any hospitalization status (32.5%), hospitalized but not admitted to the intensive care unit (ICU) (8.4%), or admitted to the ICU (2.1%).
- {gender = 'female'} with a frequency of 56,672—which reveals that 56,672 cases (i.e., 56.0% of all confirmed cases) were reported to be female.

In addition, we also found the following 1-itemset, which is of lower frequency but helps reveal additional knowledge:

- {transmission = 'not stated'} with a frequency of 7,417—which reveals that 7,417 cases were not reported with location where exposures occurred. When combining with the aforementioned discovered frequent 1-itemsets, it further reveals that community exposures accounted for 89,498 ÷ (101,121–7,417) = 95.5% of all confirmed cases with stated transmission.

Moreover, when applying vertical frequent pattern mining to these data, we also found the following frequent k-itemsets (where $k \geq 2$):

- {gender = 'female', transmission = 'domestic acquisition'} with a frequency of 50,353—which reveals that, among 101,121 confirmed cases in Canada, 50,353 cases were reported to be female transmitted through community exposures. When combining with the aforementioned discovered frequent 1-itemsets, it further reveals that a majority (i.e., 88.0%) of female cases was transmitted through community exposures. Moreover, more than half (i.e., 56.3%) of domestic acquisition cases were female.
- {death = 'no', hospital status = 'not hospitalized', transmission = 'domestic acquisition'} with a frequency of 38,464—which reveals that 38,464 cases who transmitted through community exposures survived without hospitalized.

5.2 Mining Spatio-Economic Data

We also conducted a second case study. This time, we applied vertical frequent pattern mining to find frequently "collocated" types of services in a Chinese city of Nanjing. Here, "collocation" means a collection of service providers (e.g., café, dessert house, fast food store) located within a very short walking distance within a tiny neighborhood.

By applying frequent pattern mining, we found the several frequent patterns about collocation of different service providers. Samples include:

- {fast food store, business building},
- {family restaurant, residential apartment building}, and
- {café, gym}.

These frequent patterns reveal that fast food stores tend to open within a very short walking distance of business buildings so that employees in the buildings could grasp quick lunch for the stores. Similarly, family restaurants tend to open within a very short walking distance of business buildings so that family members could enjoy their meal near home if they choose not to cook at home for that meal. Moreover, cafés tend to open within a very short walking distance of gym so that people could grasp some drinks and snacks before and after going to the gym.

6 Conclusions

Nowadays, open big data are everywhere. Data science helps discover useful knowledge and interesting information from these data. As a common data mining techniques for data science, frequent pattern mining aims to find collections or sets of frequently co-occurring patterns. In this paper, we described several vertical algorithms for mining frequent patterns from relational data. Some of them (e.g., Eclat, dEclat, VIPER) find patterns from precise data, while some other (e.g., UV-Eclat, U-VIPER) find patterns from imprecise and uncertain data. In addition, we also improved both the UV-Eclat and U-VIPER algorithms. Moreover, for illustration, we conducted two case studies. The first study focused on mining epidemiological data on COVID-19 cases in Canada, which gives researchers more insight and better understanding on the COVID-19 situations in Canada. The second study focused on spatio-economic data in Nanjing, China, which gives investors and/or city planners more insight and better understanding on the relationships and associations among different types of services within a neighborhood.

As ongoing and future work, we are exploring further improvements to vertical frequent pattern mining (e.g., the use of high performance computing). We also conduct additional studies on different big data applications and services, and evaluate the benefits of data mining to real-life applications and services.

Acknowledgements. This project is partially supported by (i) Birla Institute of Technology & Science (BITS) - Pilani, (ii) China Scholarship Council (CSC), (iii) Mitacs (Canada), (iv) Nanjing University, (v) Natural Sciences and Engineering Research Council of Canada (NSERC), (vi) Tongji University, as well as (vii) University of Manitoba.

References

1. Chen, Z., Wang, Y., Narasayya, V.R., Chaudhuri, S.: Customizable and scalable fuzzy join for big data. PVLDB **12**(12), 2106–2117 (2019). https://doi.org/10.14778/3352063.3352128
2. Lee, W., Leung, C.K. (eds.): Big Data Applications and Services 2017. AISC, vol. 770. Springer, Singapore (2019). https://doi.org/10.1007/978-981-13-0695-2
3. Leung, C.K.: Big data analysis and mining. In: Advanced Methodologies and Technologies in Network Architecture, Mobile Computing, and Data Analytics, pp. 15–27 (2019). https://doi.org/10.4018/978-1-5225-7598-6.ch002
4. Schäfer, N., Michel, S.: JODA: a vertically scalable, lightweight JSON processor for big data transformations. In: IEEE ICDE 2020, pp. 1726–1729 (2020). https://doi.org/10.1109/ICDE48307.2020.00155
5. Siddiqui, T., Jindal, A., Qiao, S., Patel, H., Le, W.: Cost models for big data query processing: learning, retrofitting, and our findings. In: ACM SIGMOD 2020, pp. 99–113 (2020). https://doi.org/10.1145/3318464.3380584
6. Leung, C.K.: Mining uncertain data. Wiley Interdisc. Rev.: Data Mining Knowl. Discovery **1**(4), 316–329 (2011). https://doi.org/10.1002/widm.31
7. Leung, C.K.-S., Mateo, M.A.F., Brajczuk, D.A.: A tree-based approach for frequent pattern mining from uncertain data. In: Washio, T., Suzuki, E., Ting, K.M., Inokuchi, A. (eds.) PAKDD 2008. LNCS (LNAI), vol. 5012, pp. 653–661. Springer, Heidelberg (2008). https://doi.org/10.1007/978-3-540-68125-0_61
8. Ma, C., Cheng, R., Lakshmanan, L.V.S., Grubenmann, T., Fang, Y., Li, X.: LINC: a motif counting algorithm for uncertain graphs. PVLDB **13**(2), 155–168 (2019). https://doi.org/10.14778/3364324.3364330
9. Leung, C.K., Zhang, Y.: An HSV-based visual analytic system for data science on music and beyond. Int. J. Art, Culture Des. Technol. (IJACDT) **8**(1), 68–83 (2019). https://doi.org/10.4018/ijacdt.2019010105
10. Martins, R., Chen, J., Chen, Y., Feng, Y., Dillig, I.: Trinity: an extensible synthesis framework for data science. PVLDB **12**(12), 1914–1917 (2019). https://doi.org/10.14778/3352063.3352098
11. Parameswaran, A.: Enabling data science for the majority. PVLDB **12**(12), 2309–2322 (2019). https://doi.org/10.14778/3352063.3352148
12. Ullman, J.D.: The battle for data science. IEEE Data Eng. Bull. **43**(2), 8–14 (2020)

13. Zhang, Y., Ives, Z.G.: Finding related tables in data lakes for interactive data science. In: ACM SIGMOD 2020, pp. 1951–1966 (2020). https://doi.org/10.1145/3318464.3389726
14. Agrawal, R., Srikant, R.: Fast algorithms for mining association rules in large databases. In: VLDB 1994, pp. 487–499. Morgan Kaufmann (1994)
15. Leung, C.K.: Frequent itemset mining with constraints. In: Liu, L, Özsu, M.T. (eds.) Encyclopedia of Database Systems, 2nd edn., pp. 1531–1536. Springer, New York (2018). https://doi.org/10.1007/978-1-4614-8265-9_17
16. Leung, C.K., Khan, Q.I.: DSTree: a tree structure for the mining of frequent sets from data streams. In: IEEE ICDM 2006, pp. 928–932 (2006). https://doi.org/10.1109/ICDM.2006.62
17. Bian, S., Guo, Q., Wang, S., Yu, J.X.: Efficient algorithms for budgeted influence maximization on massive social networks. PVLDB **13**(9), 1498–1510 (2020). https://doi.org/10.14778/3397230.3397244
18. Jiang, F., Leung, C.K., Tanbeer, S.K.: Finding popular friends in social networks. In: CGC 2012, pp. 501–508. IEEE (2012). https://doi.org/10.1109/CGC.2012.99
19. Leung, C.K.-S., Tanbeer, S.K., Cameron, J.J.: Interactive discovery of influential friends from social networks. Soc. Netw. Anal. Mining **4**(1), 154:1–154:13 (2014). https://doi.org/10.1007/s13278-014-0154-z
20. Tanbeer, S.K., Leung, C.K., Cameron, J.J.: Interactive mining of strong friends from social networks and its applications in e-commerce. JOCEC **24**(2–3), 157–173 (2014). https://doi.org/10.1080/10919392.2014.896715
21. Lee, T., Matsushima, S., Yamanishi, K.: Grafting for combinatorial binary model using frequent itemset mining. Data Mining Knowl. Discovery **34**(1), 101–123 (2020). https://doi.org/10.1007/s10618-019-00657-9
22. Leung, C.K., Zhang, H., Souza, J., Lee, W.: Scalable vertical mining for big data analytics of frequent itemsets. In: Hartmann, S., Ma, H., Hameurlain, A., Pernul, G., Wagner, R.R. (eds.) DEXA 2018. LNCS, vol. 11029, pp. 3–17. Springer, Cham (2018). https://doi.org/10.1007/978-3-319-98809-2_1
23. Zaki, M.J.: Scalable algorithms for association mining. IEEE TKDE **12**(3), 372–390 (2000). https://doi.org/10.1109/69.846291
24. Zaki, M.J.: Fast vertical mining using diffsets. In: ACM KDD 2003, pp. 326–335 (2003). https://doi.org/10.1145/956750.956788
25. Shenoy, P., Bhalotia, J.R., Bawa, M., Shah, D.: Turbo-charging vertical mining of large databases. In: ACM SIGMOD 2000, pp. 22–33 (2000). https://doi.org/10.1145/342009.335376
26. Leung, C.K.: Pattern mining for knowledge discovery. In: IDEAS 2019, pp. 34:1–34:5. ACM (2019). https://doi.org/10.1145/3331076.3331099
27. Budhia, B.P., Cuzzocrea, A., Leung, C.K.: Vertical frequent pattern mining from uncertain data. In: KES 2012. FAIA, vol. 243, pp. 1273–1282 (2012). https://doi.org/10.3233/978-1-61499-105-2-1273
28. Leung, C.K., Tanbeer, S.K., Budhia, B.P., Zacharias, L.C.: Mining probabilistic datasets vertically. In: IDEAS 2012, pp. 199–204. ACM (2012). https://doi.org/10.1145/2351476.2351500
29. Corrales-Garay, D., Ortiz-de-Urbina-Criado, M., Mora-Valentín, E.: A research agenda on open data impact process for open innovation. IEEE Access **8**, 34696–34705 (2020). https://doi.org/10.1109/ACCESS.2020.2974378
30. Leung, C.K., Chen, Y., Shang, S., Wen, Y., Hryhoruk, C.C.J., Levesque, D.L., Braun, N.A., Seth, N., Jain, P.: Data mining on open public transit data for transportation analytics during pre-COVID-19 era and COVID-19 era. In: Barolli, L., Li, K. F., Miwa, H. (eds.) INCoS 2020. AISC, vol. 1263. Springer, Cham (2020). https://doi.org/10.1007/978-3-030-57795-7_13

31. Statistics Canada: Table 13-10-0774-01 detailed preliminary information on cases of COVID-19: 6 dimensions (aggregated data). Public Health Agency of Canada (2020). https://doi.org/10.25318/1310077401-eng
32. Statistics Canada: Table 13-10-0775-01 detailed preliminary information on cases of COVID-19: 4 dimensions (aggregated data). Public Health Agency of Canada (2020). https://doi.org/10.25318/1310077501-eng
33. Statistics Canada: Table 13-10-0781-01 detailed preliminary information on confirmed cases of COVID-19 (revised). Public Health Agency of Canada (2020). https://doi.org/10.25318/1310078101-eng

Author Index

W. Lee et al. (Eds.): *Big Data Analyses, Services, and Smart Data*, AISC 899, p. 117, 2021.
https://doi.org/10.1007/978-981-15-8731-3